RAND ARROYO CENTER

T0302542

Changing the Army's Weapon Training Strategies to Meet Operational Requirements More Efficiently and Effectively

James C. Crowley, Bryan W. Hallmark, Michael G. Shanley, Jerry M. Sollinger

Prepared for the United States Army
Approved for public release; distribution unlimited

The research described in this report was sponsored by the United States Army under Contract No. W74V8H-06-C-0001.

Library of Congress Cataloging-in-Publication Data is available for this publication.

ISBN 978-0-8330-8261-9

The RAND Corporation is a nonprofit institution that helps improve policy and decisionmaking through research and analysis. RAND's publications do not necessarily reflect the opinions of its research clients and sponsors.

Support RAND—make a tax-deductible charitable contribution at www.rand.org/giving/contribute.html

RAND® is a registered trademark.

RAND OFFICES

SANTA MONICA, CA • WASHINGTON, DC

PITTSBURGH, PA • NEW ORLEANS, LA • JACKSON, MS • BOSTON, MA

CAMBRIDGE, UK • BRUSSELS, BE

www.rand.org

Preface

The ability of soldiers to engage the enemy effectively is fundamental to the operational success of the U.S. Army. As a result, the Army devotes considerable effort and resources to weapon training. The Army's current challenge is to adapt weapon training strategies to meet complex operational environments, to take full advantage of enhanced weapon training capabilities, and to be more efficient.

This document reports results of a research project entitled "Adapting Live Training Strategies to Better Support Full-Spectrum Operations and Persistent Conflict and to Leverage Advancing Technologies." The objective of the project was to support the Army's efforts to improve its weapon training strategies to better support operational requirements and unit readiness processes, take full advantage of the potential of simulator technology, and become more efficient.

This report documents the findings, conclusions, and implications of this research. It outlines findings arising from an assessment of the Army's processes for developing weapon training strategies and areas for improvement. We also examine the potential of emerging technologies to make weapon training strategies more effective and efficient. Finally, we outline directions the Army could take to improve weapon training strategies and the programs that support them, as well as to better integrate them with the broader range of Army training strategies and programs. This report should interest those involved in designing Army training strategies and those involved in the process of providing resources for implementing those strategies.

This research was sponsored by Brigadier General Michael D. Lundy, Deputy Commanding General, Combined Arms Center–Training, and conducted within RAND Arroyo Center's Manpower and Training Program. RAND Arroyo Center, part of the RAND Corporation, is a federally funded research and development center sponsored by the United States Army.

The Project Unique Identification Codes (PUICs) for the projects that produced this document are RAN126141 and HQD115815.

Contents

Figures

Tables

Summary

The ability of soldiers to engage the enemy is fundamental to operational success. As a result, the U.S. Army devotes considerable unit effort and resources to weapon training. However, the Army's leadership recognizes that weapon training strategies may need revision. The Army has broadened its operational focus to encompass a spectrum that includes conflict against nation-states, counterinsurgency, and stability operations. It has also implemented an approach to unit readiness called Army Force Generation (ARFORGEN), which moves from a concept of keeping all units constantly ready to one of progressive unit readiness. Weapon-system capabilities and the capabilities for weapon-system training have also improved. At the same time, the Army is facing substantial reductions in its budget. It is, therefore, closely examining its current resource allocations to determine where savings might be possible, including the increased use of virtual technologies in weapon training strategies.

These considerations led the Vice Chief of Staff of the Army to direct a comprehensive review of the Army's weapon training strategies. Subsequently, the Training and Doctrine Command (TRADOC) asked the Arroyo Center to support its efforts to adapt its weapon training strategies. This report summarizes results of this research, which entailed the following four tasks:

1. Examine and assess the Army's current processes for developing and adapting weapon training strategies and identify areas and directions for improvement.
2. Examine the potential of current and emerging training technologies to enhance weapon training strategies and make them more efficient.
3. Examine the implications of changing operational requirements and unit readiness processes on unit weapon training strategies and standards.
4. Draw conclusions regarding directions the Army could take to improve current weapon training strategies and provide broader implications.

Key Findings and Conclusions

Current Weapon Training Strategies and Processes for Their Development

We examined a wide range of current weapon training strategies, from small arms to helicopters and tanks. To illustrate approaches to how the Army could develop and assess the viability of alternative courses of action, we also examined various options for making rifle training more efficient. The findings from this examination are as follows:

Training strategies and standards vary by weapon system but have many similarities. Training strategies and standards outline a progressive semiannual cycle starting with preliminary training and culminating with individual/crew qualification. Most strategies also have a higher-echelon collective live-fire qualification event, but the engagement standards for these are not as well defined.

Training strategies include a wide and varied use of simulators. Simulators figured into training for all the weapon systems we examined, and over half of the training programs used simulators as a part of their qualification requirements. For weapons with high ammunition costs (e.g., the Multiple Launch Rocket System [MLRS]), training strategies were simulator based.

Training strategies and standards have not changed greatly since 2000, but training ammunition expenditure rates have. The differences between the weapon training strategies in 2000 and 2011 were small. Some changes had been made to meet new operational requirements, and there had been a small increase in the use of simulators. However, training ammunition expenditure rates had changed considerably. In 2011, these rates were around 30 percent of the authorized amounts for many types of ammunition. However, these low expenditure rates are likely not a reliable guide for determining future requirements. In 2000, before the rate of operational deployments increased, they were over 80 percent.

The wide range of resources needed to support weapon training strategies comes from many sources. These resources include ammunition, ranges, simulators, equipment, installation support, and a range of others, and they come not only from training programs, but equipping and manning programs as well.

There are multiple directions the Army could take to adjust live-fire strategies to reduce the amount of training ammunition consumed. The basic directions are: (1) reduce the number of rounds/engagements per training event, (2) reduce the number of events, or (3) reduce the frequency of events. However, experimentation and analysis is necessary to provide the Army's leadership with the data needed to make informed risk assessments.

The process for developing weapon training strategies is decentralized and evolutionary. We examined Army-level guidance on weapon training strategy and standards development processes and found that general responsibilities are outlined, but not specific guidelines. The weapon system proponents have the responsibility for developing these strategies and standards, but how their staffs do this is not standardized. Moreover, there is little understanding of operational force programs and needs, and institutional understanding of analytic bases and history is limited. The weapon training community is only staffed to support incremental change, and the mechanisms for integrating live and virtual training are informal rather than systematic.

We concluded that improved processes are needed. The Army should review each of its weapon training strategies to ensure that they make the best use of simulators, consider current unit practices, respond to changing operational needs and threats with the greatest efficiency possible, and are integrated with other training strategies. However, significant process changes will be needed to make such improvements, and this will prove challenging because there are limited staff resources to support weapon training strategy development.

Potential for Simulators to Enhance Weapon Training and Make It More Efficient

We found a consensus view that simulators have the potential to increase weapon training frequency and feedback and reduce costs. However, few studies directly showed a relationship between simulator and live-fire proficiency. In many cases, the Army seems to have assumed that a well-designed simulator would effectively train engagement skills. We also found that many obstacles limit the expanded the use of simulators. Many weapon-system strategies already include the extensive use of simulators, and expanding simulator capabilities requires time and investment prior to achieving any benefit. From these findings, we drew the following conclusions:

Simulators have the *potential* to improve weapon training. This includes increasing training frequency and feedback, training tasks under conditions not possible in live training, and reducing costs. The question of whether these improvements are achievable should be examined before acquiring additional, improved, or new simulators.

A training effectiveness analysis should be done early to ensure the development of a system that can yield real training benefit at a reasonable cost.

It is uncertain whether increased simulator use will significantly reduce training ammunition costs. Simulators currently play a major role in training strategies, the high-value targets (weapon systems with high-cost ammunition) are gone, and it can take time and up-front resources to acquire new, improved, or additional simulators.

Implications of Changing Operational Requirements and Unit Readiness Processes

We examined small-arms weapon strategies in detail to determine how well they reflect the engagement skills needed for combat success and how well they support the achievement of these standards in the context of ARFORGEN readiness processes. The following key findings arose from this examination:

- Many critical individual/crew small-arms tasks and skills are not included in qualification standards or fully supported in the strategies.
- Most of the individual/crew strategy training ammunition is allocated to repetitive qualification events, but little is allocated to preparation events and activities that develop basic marksmanship skills.
- Collective (above crew) live-fire exercises account for a large percentage of training ammunition and other resources but have no engagement standards. Also, there is not enough range capability for the required collective events.
- While there are not enough fielded simulators to support small-arms training strategy and qualification requirements, simulator usage is low.
- There is limited structured training of small-arms "train the trainer" skills.
- Unlike in the U.S. Marine Corps (USMC) and the armies of many nations, there is no Military Occupational Specialty (MOS) for Army machine gunners.

These findings suggest that standards should include a broader range of needed combat engagement skills and that the strategies should change to include increased resources (e.g., ammunition and simulators) for preparation and to align better with progressive ARFORGEN readiness processes.

Specific directions the Army could take to improve these standards and strategies include

- develop enhanced, prescriptive individual and crew standards using the full range of virtual and live-fire modalities
- improve the benefits of collective live-fire exercises

- develop descriptive "resource neutral" strategies to achieve combat standards aligned to support progressive ARFORGEN readiness goals
- identify, develop, and field cost-effective training capabilities
- develop and institute formal "train the trainer" programs.

Implementation should start in the near term by adding critical combat tasks/skills as qualification requirements. While identifying the needed tasks/skills and developing descriptive small-arms preparation strategies could be done based on military judgment, defining the standards would require experimental studies to validate the standards and confirm hypotheses, such as whether simulator performance correlates to live-fire performance.

Longer-term efforts should refine and adjust strategies as emerging operational requirements and the training environment become more defined and unit commanders shape their training programs to meet new requirements and constrains. The Army should set up a program to monitor and understand how unit programs evolve, how well they reach prescribed small-arms standards, how well they align with small-arms strategies (and, if they differ, why), and in what ways they could be improved and better supported. Structured observational studies should support analysis and improvement.

Improving the Broader Range of Weapon Training Strategies

Based on our focused examination of small-arms training strategies, coupled with our examination of the broader range of weapon systems, we have identified a set of general conclusions and directions for improvement.

Conclusions

Training strategies and the resources to support them should vary by weapon system. Each weapon system is different; therefore the training strategy for each needs to be shaped and supported differently.

While weapon training strategies may be improved by changing the mix of live-fire and virtual modalities, such improvement will require effort and study. The increased use of simulators to train tasks not feasible using live-fire modalities would be a valuable supplement to current strategies and could increase efficiency. This potential is likely to increase as weapon and simulator capabilities improve. However, making such improvements will prove challenging, as each weapon system requires close examination, tests, and studies to underpin change. Also, the staff resources are limited, simulators are already used extensively, and adding simulator capability requires up-front time and resources.

It is important to identify and verify combat-critical weapon engagement tasks and skills. Considerable qualitative military judgment would be required and determining the minimum level of engagement skills should be an Army-level decision.

Better alignment with ARFORGEN readiness goals and manning strategies could achieve needed readiness levels more efficiently. Preparation activities could take place during the first half of the ARFORGEN process, individual and crew qualification could occur toward the middle, and collective live-fire qualification could happen near the end of the first year. The number of qualification events would be reduced and better integration with ARFORGEN manning strategies could be achieved.

A lack of data undercuts the Army's ability to make needed improvements. The Army does not have a system for knowing how closely units follow current weapon training strategies, the degree to which they achieve standards, or how well current standards relate to combat success.

It is important to coordinate weapon training strategies to ensure needed resources are available. Weapon training resources come from many different sources, and ensuring that changes to weapon training strategies are feasible and can be supported is important.

Further definition of the processes for developing and reviewing weapon training strategies would enhance improvement. Weapon training strategies could benefit from a review similar to the one done for small arms. It is important to be proactive to avoid potentially damaging across-the-board cuts when more targeted reductions are possible. A more basic issue is that the Army has not defined an overarching process that integrates weapon training strategies with other training strategy components.

Directions for Improving Weapon Training Strategies

Given these conclusions, we identify three directions the Army should consider:

1. **Develop definitive guidance on weapon training development processes.** Key elements of this guidance should be requirements to identify critical engagement tasks and standards, achieve critical tasks at the lowest cost possible, optimize the mix of training modalities, validate strategies before implementation, coordinate weapon training strategies with other training strategies, and continuously review and assess strategies.
2. **Collect and analyze data on unit weapon training programs and proficiency levels.** Many key elements needed for improving training strategy are not available.
3. **Develop and improve weapon training strategies as part of a larger, integrated process.** Changes in one area affect others and all use a common set of resources. Likewise, collecting data on unit training activities would be more efficient if done as a coordinated effort. Revised training strategies should define a feasible balance of training activities needed to meet changed operational requirements. This suggests that some elements may have to grow and others be reduced or eliminated.

Broader Implications

A recently published RAND Arroyo Center report examining the broader range of the Army Training and Leader Development (ATLD) programs presented an overall conclusion that current ATLD management processes were developed for incremental change, while fundamental changes are required.[1] More understanding and focus on operational-force needs, increased integration between ATLD and other programs, and a more structured cost-benefit approach were identified as areas for improvement. The weapon training research we document in this report supports this overall conclusion and the need for improvement in these areas. However,

[1] This research is documented in James C. Crowley, Michael G. Shanley, Jeff Rothenberg, and Jerry M. Sollinger, *Adapting the Army's Training and Leader Development Programs for Future Challenges,* Santa Monica, Calif.: RAND Corporation, TR-1236-A, 2012.

we also posit the need to ensure that training strategies address all tasks and skills that are critical to operational success.

The earlier research also outlined a number of specific directions for improvement. Our weapon training research directly supports the potential benefits of most of these directions and strongly supports the concept of forming a single organization with overall responsibility for data collection, analysis, and support of ATLD program management. Another finding from the earlier report that is amplified by our research is the need for better integration among the training and other programs.

Acknowledgments

This two-year research effort was initially sponsored by Brigadier General Paul Funk, then by Brigadier General Robert "Pat" White, and finally by Brigadier General Michael D. Lundy, Deputy Commanding Generals for Training, U.S. Army Combined Arms Center–Training (CAC-T). Without their support and that of their staff members, this study would not have been possible. At CAC-T, Denny Tighe, Bob Banning, Kent Ervin, and Harold Summerfeld also provided great assistance.

From the Army Training Support Center, George Baker was invaluable as our primary contact and coordination point for necessary Army and Marine Corps support. Mark Mueller, Ed Grinnell, Bill Andersen, Darryl Powell, and Doug Prior also provided valuable support.

We are also indebted to Colonel Patrick Connor, Glenn VonSwol, Ray Ingram, Tom Scarbath, Ira Partridge, Matthew Golden, and Dennis Terry at the TRADOC Capability Manager Live, as well as Dan Wakeman, Tim Sayers, and Devon Bradley at the TRADOC Capability Manager Virtual.

Achieving the study's goals required an in-depth understanding of weapon training strategies and the factors that underpin them. The high level of support we received from the subject-matter experts at TRADOC's Centers of Excellence was key to our efforts. Specifically providing this support were Bill Lodes, Clantford Drummond, Dan Etheridge, Steve Sells, Billy Payne, Houston Lesley, Alan Page, and Don Schlomer at the Fires Center of Excellence; Chief Warrant Officer Robert Jackson and Debra Parks at the Aviation Center of Excellence; Traci Kuehl at the Maneuver Support Center of Excellence; and Mark Girard at the Sustainment Center of Excellence.

A focus of our research effort was small-arms training strategies, and we had exceptional support from the Maneuver Center of Excellence. Colonel Emmett Schaill, Lieutenant Colonels Brad Nadig and Dan Hodne, Major Sean Vinson, Sergeants First Class Tony Frazer and Don Mullins, and Staff Sergeants Josh Tyler and Tim Billue, as well as Allen House, Steve Krivitsky, Harold Lusano, Mike Wood, Rich Funke, Jurgen Becker, John McDowell, and the Army Research Institute's (ARI's) Jean Dyer, were very generous in their support. A special thanks is due to Captain Dan Wilcox, whose efforts to move "combat shooter" skills to the forefront will provide long-term benefits to the Army. At Fort Eustis, Mark Leclair's demonstration of the capabilities of the Army's small-arms simulator gave us an important hands-on understanding of its capabilities. Finally, special acknowledgment must be given to Fred Roberts, who was tragically killed during the course of this research effort. He was extremely helpful to our research team, but more importantly his efforts in support of the Army's training mission were large and significant. He will be missed.

An important part of our small-arms training research involved the examination of the small-arms training programs of other nations and services. In this effort, we received valuable support from Sergeant Majors Peter Gould and Jack Nevils at the 1st Special Warfare Training Group, Fort Bragg, and from Chief Warrant Officer Martin "Dank" Dankanich from the Marine Corps' Weapons Training Battalion. Lieutenant Colonel Youngsub Lee of the Republic of Korea Army, Lieutenant Colonel Philippe Testart of the French Army, Lieutenant Colonel Michael Ahern and Major Paul Hopkins of the Australian Army, Colonel Guy Wallis and Major Kevin Sandison of the British Army, and Lieutenant Colonel Dan Jakubiec of the Canadian Army all generously provided background materials and insights on their army's programs.

Elsewhere within the U.S. Army, we received considerable support from Bob Grubbs, Mark Cassel, and Willie Rice in the Department of the Army's Directorate of Training. Special thanks are also due to Tim Barnhart, David Bridier, Derek Flood, and Russ Hummel at the U.S. Army Forces Command.

This work also benefited from the support of Michael Hansen, Pete Schirmer, and Chip Leonard at RAND. Major Tim Vail served as an important member of the project team while serving as a RAND Army Fellow. Finally, the thorough and thoughtful reviews of Dr. Neil Carey at CNA and John Ausink at RAND were very helpful in the process of refining and finalizing this report.

We have benefited greatly from assistance provided by all these individuals. Errors of fact or interpretation, of course, remain the authors' responsibility.

Abbreviations

AC	Active Component
AFATDS	Advance Field Artillery Tactical Data System
ARFORGEN	Army Force Generation
AR	Army Regulation
ARI	Army Research Institute
ARM	advanced rifle marksmanship
ARNG	Army National Guard
ARRM	Army Range Requirements Model
ASCC	Army Service Component Command
ASI	Additional Skill Identifier
ATLD	Army Training and Leader Development
BCT	Brigade Combat Team
BFV	Bradley Fighting Vehicle
CAC-T	Combined Arms Center–Training
CALFEX	combined arms live-fire exercise
CBRN	chemical, biological, radiological, and nuclear
CCMD	combatant command
CCO	Close Combat Optic
CCTT	Close Combat Tactical Trainer
CEF	Contingency Expeditionary Force
COE	Center of Excellence
COFT	Conduct of Fire Trainer
CTC	Combat Training Center

DA	Department of the Army
DEF	Deployment Expeditionary Force
EST	Engagement Skills Trainer
FDC	Fire Direction Center
FORSCOM	U.S. Army Forces Command
FSCATT	Fire Support Combined Arms Tactical Trainer
KD	known distance
LCT	Longbow Crew Trainer
LFX	live-fire exercise
LMTS	Laser Marksmanship Training System
LOHAM	Location of Hit and Miss
MACS	Multipurpose Arcade Combat Simulator
MDEP	Management Decision Evaluation Package
MGS	Mobile Gun System
MLRS	Multiple Launch Rocket System
MOS	Military Occupational Specialty
MTP	Mission Training Plan
NCO	noncommissioned officer
PEG	Program Evaluation Group
PMI	Preliminary Marksmanship Instruction
RC	Reserve Component
SAW	Squad Automatic Weapon
SDM	Squad Designated Marksman
SRTA	Short Range Training Ammunition
STRAC	Standards in Training Commission
TAMIS	Total Ammunition Management Information System
TC	Training Circular
TEA	Training Effectiveness Analysis
TEO	Training and Evaluation Outlines
TRADOC	Training and Doctrine Command

TWS	thermal weapon sight
USMC	U.S. Marine Corps

CHAPTER ONE

Introduction and Background

The ability of soldiers and crews to engage the enemy successfully has been fundamental to the operational success of the U.S. Army since its founding and remains so today. As a result, the Army devotes considerable resources and effort to developing and maintaining weapon-system proficiency. In 2010, over a billion dollars were spent to buy training ammunition, and many dollars were spent maintaining and providing the training aids, devices, simulators, simulations, ranges, and targetry (TADSS-RT) that support weapon-system training.[1]

Additionally, units have traditionally devoted considerable time to weapon-system proficiency. In 2001 and 2002, combat brigades devoted about 45 days—half of their annual field training days—to live- and dry-fire exercises, and they spent additional time on preliminary weapon-system training in garrison.[2]

Since 2002, the Army has broadened its operational focus and has implemented a new Army Force Generation (ARFORGEN) approach to unit readiness. These shifts have caused the Army to consider options for achieving better alignment among the weapon training strategies outlined in DA Pamphlet 350-38, *Standards in Training Commission (STRAC)*; broadened operational requirements; and ARFORGEN readiness strategies. It is also examining options for more effectively using weapon-system simulators both to enhance training and use scarce training resources more efficiently.[3]

These considerations led the Vice Chief of Staff of the Army to direct the Army's Training and Doctrine Command (TRADOC) to "evaluate the appropriate mix of live-fire and

---

[1] Training ammunition cost data provided by the Department of the Army (DA) G3/5/7 training staff. Cost of the other elements is also large. For example, the cost of maintaining ranges is over $300 million annually. We did not try to more precisely quantify the dollars in all the other weapon training elements, as this would have required a major effort. While ammunition dollars come from a single account, the dollars for the other elements come from many different accounts, and some of these accounts support more than just weapon training.

[2] This was about half of the days spent in field training. See Michael Shanley, James C. Crowley, Matthew W. Lewis, Ralph Masi, Kristin J. Leuschner, Susan G. Straus, and Jeffrey Angers, *Supporting Training Strategies for Brigade Combat Teams Using Future Combat Systems (FCS) Technologies*, Santa Monica, Calif.: RAND Corporation, MG-538-A, 2007.

[3] The terms *virtual* and *simulator* overlap somewhat. The *DoD Modeling and Simulation (M&S) Glossary*, March 2010, defines *virtual simulation* as involving "real people operating simulated systems." It defines a *training simulator* as a "device which duplicates the essential features of a task situation and provides for direct human operation," and it defines *Virtual Training Domain* as "[a] simulator-based training environment that trains real people using virtual simulators that physically replicate the working environments of real mission systems operating within realistically simulated operational battlespace environments and scenarios." In this report, we use the term *weapon-system simulator* to refer to devices that physically replicate the operation of the weapon system, enabling individuals and crews to practice using actual weapon-system engagement in a live-fire mode, and that give accurate feedback on engagement success (e.g., where a round would have landed if a real weapon with live ammunition had been used).

virtual training simulators to supplement or replace live-fire collective and individual tasks."[4] As a result, TRADOC initiated a major review of weapon training strategies. Subsequently, TRADOC's Deputy Commanding General, Combined Arms Center–Training (CAC-T), requested RAND Arroyo Center support for its review.

Purpose

The objective of this research project was to support the Army's efforts to adapt its weapon training standards and strategies to meet a broader spectrum of operational requirements more effectively, to improve alignment with ARFORGEN processes, to take full advantage of the potential of simulators, and to become more efficient. Training standards specify how well and how frequently an individual, crew, or firing unit must perform a specific weapon-system training activity to be officially considered qualified. Training strategies specify the doctrinal way individuals, crews, and firing units should train on the weapon system. Strategies include standards.[5]

Achieving this study's objective involved accomplishing the following four research tasks:

1. Examine and assess the Army's current processes for developing and adapting weapon training strategies and standards to identify areas and directions for improvement.
2. Examine the potential of current and emerging weapon-system simulators and other technologies to enhance weapon training strategies and make them more efficient.
3. Examine the implications of changing operational requirements and unit readiness processes on unit weapon training strategies and standards.
4. Draw conclusions regarding directions the Army could take to improve current weapon training strategies and present broader implications.

This report documents the findings, conclusions, and implications of this research.

Operational Requirements and ARFORGEN Readiness Processes

This research supports the Army's efforts to improve its weapon training strategies given changing operational requirements and readiness processes. This section summarizes some of the key directions of these changes.

Operational Requirements

Before 2002, unit training programs and weapon training strategies could focus on the defeat of conventional enemy forces. While this mission remains, the current operational environment requires that emphasis also be placed on defeating unconventional enemy forces using asymmetric tactics, protecting and supporting local populations, and supporting some civil

---

4 See Department of the Army, "Training Ammunition & Simulation VCSA Portfolio Review," briefing, May 2010b.

5 From this perspective, standards are "prescriptive" and strategies are "descriptive."

functions.[6] Thus, the Army must be prepared to face a wide range of threats to meet its current security requirements.

Coupled with this broader range of missions is a large increase in the operational capabilities of U.S. Army and joint forces. This combination of changing and uncertain threats and increased operational capabilities generates a need to reconsider weapon training strategies and standards, and this means that the frequency and relative importance of certain types of weapon engagements and the conditions under which they are executed could change.

Weapon training strategies and standards necessarily involve a compromise. Unit training time, ammunition, simulator capabilities, ranges, and targetry are constrained resources. As a result, weapon training strategies and standards are shaped to gain the highest level of operational weapon engagement proficiency practically possible, rather than the ideal level. As operational requirements change, a relook at these compromises is certainly warranted, even if the result is confirmation.

ARFORGEN Readiness Processes

To meet the operational demands that arose after September 11, 2001, the Army developed and implemented a new unit readiness process called Army Force Generation.[7] The goal of ARFORGEN is to provide a continuous output of ready expeditionary forces both to meet ongoing and planned operational requirements and to handle unplanned operational requirements. Under ARFORGEN, units cycle through three pools, each connoting a progressively higher state of readiness: Reset, Train/Ready, and Available.

Reset: Units enter this pool upon return from deployment or after one year in the Available pool. The intent of Reset is to restore a unit's personnel and equipment to the level needed for collective training. During the Reset period, units are assigned missions. Some will be designated Deployment Expeditionary Force (DEF) units and will prepare to deploy to support ongoing operational requirements. Others will be designated Contingency Expeditionary Force (CEF) units and will train for planned or unexpected contingencies.

Train/Ready: Units in the Train/Ready pool continually increase training readiness, performing individual and collective training to gain DEF or CEF required readiness levels.

Available: DEF units deploy to meet ongoing operational requirements and CEF units train to sustain readiness levels and deploy if required.

The key implication ARFORGEN has for weapon training strategy is the need to shift from a steady-state concept to one of progressive readiness.

To support the achievement of progressive unit readiness and mitigate risk, the Army has established ARFORGEN Aim Points. Aim Points are manning, equipment, training, and readiness targets specified at points in time along the ARFORGEN cycle to support the synchronization of manning and equipping capabilities with training. The Aim Points and associated standards are not fixed. Rather, they adjust continually as Combatant Command

---

[6] The requirement for the Army to conduct operations across a broad range of possible missions and operational environments is codified in Army Doctrinal Publication (ADP) 3.0, *Unified Land Operations,* September 2011. See also TRADOC Pamphlet 525-3-1, *The United States Army Operating Concept*, August 2010 for a general outline of the strategic and operational contexts within which the Army expects to operate and the character of the threat.

[7] The ARFORGEN purpose, policies, and responsibilities are prescribed in Army Regulation (AR) 525-29, *Army Force Generation*, Washington, D.C., March 2011b.

(CCMD) and Army Service Component Command (ASCC) requirements and Army resource availability change.

Equipping, sustaining, and manning readiness have a major influence on unit weapon training programs. Effective weapon training requires enough operational weapon systems to allow the training to be conducted, and personnel fill and stability are important for training effectiveness and efficiency. Crew and other collective live-fire training require reasonable fill rates to begin. Personnel turnover directly affects the need to retrain crews and individuals assigned to man designated weapons.[8]

The Army is now in the process of revising the ARFORGEN time lines, reducing the Active Component (AC) cycle time from 36 to 24–27 months. This change has the potential to provide for greater personnel stability throughout the entire cycle. Another goal is to align specific units with a specific CCMD/ASCC to allow planning and training activities throughout the ARFORGEN cycle to be better focused.

Research Approach

The initial step in our research was to develop a general understanding of current weapon training strategies and their standards. To do this, we examined the training strategies of 14 weapon systems. These represented a range of important weapon types from the Army's three main proponents responsible for developing weapon training strategies. This examination included the use of weapon system simulators and other technology in these strategies. It also included an examination of the degree to which units actually use the training ammunition authorized in DA Pamphlet 350-38, *Standards in Weapons Training.*

We next examined the potential of simulators to provide increased support to weapon system training strategies and the processes being used to develop and adapt these strategies. Based on the findings from these examinations, we drew conclusions about the potential of simulators to enhance and make these strategies more efficient and about the areas in which the Army's processes for developing these strategies need improvement.

To accomplish our third task, we started with a more focused examination of a smaller set of weapon systems in the small-arms category. We chose small arms for two reasons:

1. All soldiers use small arms, thus they are highly important to operational success and force protection.
2. Small-arms ammunition represents a large portion of the Army's annual training ammunition budget.

We examined a smaller set of weapon systems because a more detailed examination and understanding was needed than could be accomplished across the larger set.

In this examination, we compared the doctrinal use of those weapons with qualification standards to determine the degree to which the strategies support current operational require-

[8] Some weapons do not have individuals or crews assigned to man them as their main duty. These are called designated weapons. For example, a maintenance company may have some machine guns authorized for self-protection that are manned by soldiers with other principal duties.

ments. We also compared the strategies with unit readiness processes. Another aspect of this research involved the capabilities of simulators and ranges to support small-arms strategies.

Based on the findings of this examination, we identified areas where small-arms strategies should be improved. We also identified directions the Army could consider taking to improve small-arms training. In addition, to gain a better perspective on how the Army might adapt its strategies, we provided an illustrative approach to show how the Army could develop, study, compare, and validate options for improving weapon training strategies.

Based on our examinations of both a broad and a focused set of weapon systems, we developed a general set of conclusions about weapon system training, as well as directions the Army could take for improvement.

Finally, based on this and other research, we developed a broader set of conclusions and implications for improving the Army's processes for developing and supporting its larger set of training strategies and programs.

Research Activities

Our research included a review of documents specifying the current weapon training strategies and standards. The Army's key document in this regard is DA Pamphlet 350-38, *Standards in Training Commission (STRAC)*. This pamphlet outlines the events involved in weapon training and identifies the specific events required for qualification. It also states how often the qualification and strategy events are to be executed and the ammunition authorized for each event.

DA Pamphlet 350-38 outlines the general framework for weapon training but not the details. The specifics of weapon training activities and the standards that apply are documented in the weapon system–specific field manuals and training circulars that underpin this pamphlet. These supporting documents also contain other strategy events that are not outlined in the STRAC pamphlet.

We also reviewed documents that further describe the operational employment of weapon systems and the Army's processes for developing weapon training strategies and standards.[9] We also reviewed a wide range of documents that discuss the potential of simulators and other technologies to support training more broadly and to support weapon system training specifically.

After conducting this document review, we met with staff members involved in weapon employment and training strategies and their support from Headquarters DA and TRADOC's Combined Arms Center, Army Training Support Center, and its live and virtual capability managers and weapon system training proponents at its various Centers of Excellence, especially the Maneuver Center of Excellence (COE), which is the proponent for small-arms training and doctrine.[10] We did this to increase our understanding of

- what the strategies and standards are
- the processes for developing and supporting the strategies and standards
- how the strategies and standards are validated
- when a major review was last conducted

---

[9] In addition to DA Pamphlet 350-38, we reviewed AR 350-1, *Training and Leader Development*, August 2011; AR5-18, *Total Army Munitions Requirements Process and Prioritization System*, December 2009; and TRADOC Pamphlet 350-70-1, *Guide for Developing Collective Training Products*, May 2004.

[10] Many, but not all, of these training community members are listed in this report's Acknowledgments section.

- how the strategies and standards are adapted for new requirements
- what factors determine the use of simulators and other technologies in these strategies.

Subsequently, we vetted our findings with these contacts to ensure that they were factually accurate and that our conclusions were feasible.

Organization of the Remainder of This Report

In Chapter Two, we present our findings regarding the nature of current weapon training strategies and standards and our findings and conclusions about the Army's current processes for developing and adapting weapon training strategies.

Chapter Three outlines our findings and conclusions regarding the potential of simulators and other technologies to enhance current weapon training strategies and make them more efficient.

Chapter Four presents the results of our more focused examination of small-arms training standards, strategies, and their support.

Chapter Five identifies areas where small-arms strategies could be changed to provide better support to operational requirements and unit readiness processes. It also identifies directions that the Army could consider for improving small-arms weapon strategies.

In Chapter Six, we outline an approach for developing, comparing, and validating feasible small-arms training strategy options.

In Chapter Seven, the first section presents the overall conclusions we drew from both our examination of small-arms training strategies and the full range of weapon training strategies outlined in Chapter Two. In the second section, we outline directions the Army could take to improve its processes for developing and improving its weapon training strategies. In the third section, we outline the implications this research has for the Army Training and Leader Development programs.

CHAPTER TWO

Weapon Training Strategies and Processes for Their Development

In this chapter, we first outline the findings that arose from our examination of the training strategies of 14 weapon systems. We then present our findings and conclusions on the current state of processes for developing weapon training strategies, standards, and support.

The Current State of Weapon Training Strategies and Standards

We examined the strategies and standards for the 14 weapon systems shown in Table 2.1.

These systems represent a wide range of individual and crew weapon system types and account for a large percentage of the Army's annual training ammunition expenditures. The proponents for these weapon systems are three of TRADOC's five Centers of Excellence. Additionally, while the other two centers—the Sustainment COE and the Maneuver Support

Table 2.1
Weapon Systems Examined

Proponent	Weapon System
Maneuver COE	M4/16 Rifle
	M9 Pistol
	M2, MK19, and M240B Machine Guns, M249 Squad Automatic Rifle
	Bradley Fighting Vehicle
	Abrams Tank
Fires COE	155mm and 105mm Howitzer
	Multiple Launch Rocket System (MLRS)
	Avenger Air Defense System
Aviation COE	AH64 Attack Helicopter
	OH58 Observation Helicopter

SOURCES: FM 3-22.9, *Rifle Marksmanship M16/M4-Series Weapons*, February 2011a; FM 3-23.35, *Combat Training with Pistols, M9 and M11*, June 2003; FM 3-22.27, *MK 19, 40mm Grenade Machine Gun, MOD3*, November 2003; FM 3-22.68, *Crew Served Machine Guns 5.56-mm and 7.62-mm*, July 2006b; FM 3-20.21/MCWP 3-12.2, *Heavy Brigade Combat Team (HBCT) Gunnery*, September 2009a; TC 3-09.8, *Field Artillery Gunnery (Revised Final Draft)*, November 2010; TC 3-01.63, *Avenger Gunnery Program (Initial Draft)*, May 2011; and FM 3-04.140(FM 1-140), *Helicopter Gunnery*, July 2003.

COE—are responsible for developing their own small-arms training strategies and standards, in practice they use the strategies and standards developed by the Maneuver COE.[1]

Weapon Strategies and Standards Vary by Weapon System but Have Many Similarities

As would be expected, specific weapon systems' strategies vary in terms of number and types of events, standards, and the use of simulators and other technologies. However, the strategies share a similar structure in many respects.

First, all strategies outline a progressive set of training activities and events, most of which are captured in a set of defined dry- and live-fire "Tables."[2] Across the weapon systems we examined, training begins with individual training and culminates with collective live-fire exercises. Individual training starts with training on weapon system operation and engagement techniques and normally includes a "gunnery skills test" or some other type of hands-on test of equipment operating skills. These strategies then progress to dry-fire exercises—sometimes supported by simulators or gunnery devices—and then to individual or crew weapon qualifications, with most weapon systems having a live-fire practice before the individual or crew live-fire qualification event. All these strategies include above crew–level collective live-fire events. Except for helicopters, pistols, and some support units, successful collective live-fire exercises are required for AC qualification.[3]

Aside from helicopter gunnery, which occurs annually, the AC gunnery strategies involve going through the cycle of individual and crew training activities and practice and qualification events every six months. For Reserve Component (RC) units, the cycle is 12 months.

For some systems, collective live-fire exercises are not part of the six-month cycle, but are conducted annually. For example, tanks and Bradley Fighting Vehicles (BFVs) have platoon-level live-fire qualification and the company or above combined arms live-fire exercises (CALFEXs) annually.[4]

RC strategies have reduced collective live-fire events. For example, RC tank-unit strategies do not include a platoon live-fire qualification or CALFEX.[5]

Standards for individual and crew qualification are specified in terms of accuracy (hits relative to the number of rounds fired) and the time required to get a hit. Target types and ranges are specified for direct-fire weapon systems, and most have requirements for engagements under chemical, biological, radiological, and nuclear (CBRN) conditions and, if issued, with night-vision equipment.

Most strategies allow for requalification firing if the standards are not met the first time; however, DA Pamphlet 350-38 does not specifically authorize ammunition for such training.

---

[1] There are some exceptions. For example, the Maneuver Support COE develops specific M9 pistol strategies and standards for Military Police soldiers.

[2] In weapon training strategies, many structured training events are called tables.

[3] For the remainder of this report, *collective training* refers to organizations above crew level. Technically, a crew is a collective organization but it mans a single weapon system, so we consider crew and individual weapon training as a different category from collective training. In this report, we use the term *collective training* to refer to training an organization that uses multiple weapon systems, such as an infantry squad or platoon.

[4] Also, infantry units have a required squad/platoon-level live-fire exercise (LFX) qualification quarterly.

[5] This is for pre-mobilization strategies. After being mobilized and preparing for a combat operational deployment, collective live-fire events would normally be a part of the training program.

Standards are less specific for platoon and above collective live-fire qualification events, with commanders given much latitude to shape the events to meet their mission needs and to determine qualification criteria.

The Use of Weapon-System Simulators Is Varied but Extensive

An examination of current strategies shows extensive simulator use in weapon training strategies, but this use varies by the weapon system's training needs, training ammunition costs, and training simulator capabilities.

Simulators play the largest training role for weapon systems with high per-round costs. Key examples are anti-tank and air defense guided missile systems (including the tube-launched, optically tracked, wire-guided [TOW] missile, Javelin, Patriot, and Hellfire), for which training and qualification is simulator based and live fire is not normally part of the strategy.

For other weapon systems, simulator training is the main part of the strategy, including the Multiple Launch Rocket System (MLRS), Apache, and Avenger. While there is some live fire, most training is technology based, using either embedded trainers or stand-alone simulators.[6]

In the case of other weapon systems, strategy and qualification is primarily live fire, but simulators are part of the training strategy and are required for qualification. This is the case for machine gun 10-meter practice and qualification fire and for CBRN and unassisted night-fire practice and qualification for M4/M16 rifles and the M9 pistol.

For some weapon systems, simulator events are a "gate" in that the crew must successfully complete simulator-based training to go on to the live-fire events. The most prominent example is Conduct of Fire Trainers (COFTs) training for the tank and BFV. Another example is forward observer (FO) training on Call for Fire Trainers and fire direction center section training on an Advanced Field Artillery Tactical Data System (AFATDS) simulator/stimulator for the Paladin. Simulators also play roles in preliminary training for most weapon systems, especially small arms.

Simulators have a role in the training strategies of all 14 of the weapon systems we examined: eight use simulators/embedded gunnery devices as a part of their qualification requirements, four have simulator training as gates, and two use simulators as part of their training strategies.

Strategies and Standards Have Not Changed Greatly Since 2000, but Training Ammunition Expenditure Rates Have Changed

We compared documents outlining weapon training strategies in 2000 with those for 2011 for the M4/M16 rifle, M240 machine gun, tank, BFV, Paladin 155mm Howitzer, and MLRS.[7] We found that while there had been some changes in terms of training ammunition require-

---

[6] Embedded training is where the simulator is actually a part of the operational equipment. The examples among the weapon systems we examined are the Apache and MLRS.

[7] We used a 1998 draft DA Pamphlet 350-38 to determine the FY 2000 training ammunition requirements; the DA and TRADOC STRAC staff did not have copies of the approved pamphlet, and we could locate only draft versions. For 2011, we used the version of the DA Pamphlet being finalized in 2011, which was provided to us by the TRADOC Capability Manager's STRAC division.

ments and the use of simulators, overall differences were small[8] and did not reflect the change in the Army's readiness approach to an ARFORGEN cycle, as these strategies still outline live-fire training on a semiannual cycle rather than a progressive readiness approach as the unit approaches the date it is available for deployment.

The major changes in annual live-fire training ammunition requirements for AC soldiers are shown on Table 2.2, which compares the requirements from 2000 with those from 2011.

Table 2.2
Training Ammunition Requirement Changes, FY 2000–2011 (AC)

Weapon	FY 2000 Rounds per Weapon	FY 2011 Rounds per Weapon	Difference	Remarks
M16/4 Rifle Infantry	Individual: 739	Individual: 798	–22	Engagement Skills Trainer (EST) replaced live fire for unassisted night and CBRN
	Collective: 1,110	Collective: 1,010		Fewer live-fire exercises (LFXs)
	Total: 1,840	Total: 1,818		Night-sight practice and qualification added
M16/4 Non-infantry	Individual: 198	Individual: 196	+128	Changed to semiannual qualification
	Collective: 0	Collective: 130		LFX added
	Total: 198	Total: 326		EST replaced live fire for unassisted night and CBRN
M240B Machine Gun Infantry	Crew: 2,304	Crew: 2,004	–300	EST for 10-meter
	Collective: 2,000	Collective: 2000		Thermal expanded
	Total: 4,304	Total: 4,004		
Paladin Field Artillery	198	160	–38	Reason unknown
MLRS Field Artillery	6	9	+3	Reason unknown
M1 Tank Armor	MG: 104	MG: 102	MG –2	Added MG engagements for changed threat
	Sub-cal: 58	Sub-cal: 81	Sub-cal: +23	Section qualification added
	50 cal: 1,218	50 cal: 1,850	50 cal: +532	
	7.62: 2,863	7.62: 5,400	7.62: +2,337	

NOTE: Cal = caliber; MG = machine gun.

[8] The training ammunition requirements listed do not include blank ammunition, which is used primarily for tactical collective training.

Major changes included increases in rifle training ammunition for non-infantry soldiers and machine gun ammunition for tank crews, as well as a reduction in ammunition for M240B machine gun crew training. The increase in rifle ammunition reflected the fact that qualification for non-infantry soldiers had become a semiannual rather than an annual requirement, as well as the addition of a collective live-fire exercise (LFX) for non-infantry soldiers. The reason for this change was the recognition that small-arms self-protection is increasingly important given the operational environments in Iraq and Afghanistan and other likely future conflicts. These increases were somewhat offset by the use of a small-arms simulator—the Engagement Skills Trainer (EST)—for unassisted night fire and CBRN qualification for all soldiers with access to an EST.

Tank machine gun training ammunition requirements rose as the need for improved skills in engaging dismounted enemy soldiers was identified as an important operational requirement, and the number of machine-gun engagements in the tank tables increased.

The reduction in live-fire training ammunition requirements for infantry M240B machine guns was caused by a shift to the EST for 10-meter fire and a reduction in the number of rounds authorized for LFXs.

The number of rounds for field artillery went down (from 198 to 160) for Paladin and increased for MLRS. We could not determine the reason for these changes. For the Paladin, it is possible that there was recognition that improved equipment, fire direction software, and position location capabilities inherently improved accuracy, thus reducing the need for using as many live-fire rounds, but we could not establish whether improved Paladin system capability was the reason behind the reduction.

While some changes were made due to the increased use of simulators—specifically the EST—and due to changed operational requirements and doctrine, the structure of weapon training strategies and overall training ammunition requirements changed only slightly.

In terms of simulator use, there were no added capabilities, but the use of the EST increased.[9]

Training Ammunition Expenditure Rates. A key objective of this study effort was to support the Army in identifying options for making its weapon training strategies more efficient, especially in terms of ammunition use. For this reason, we next compare requirements (the quantity of training ammunition stated in DA Pamphlet 350-38 that a unit must have to execute its training mission) and authorizations (the amount of ammunition a unit commander can draw in a fiscal year) with actual unit training ammunition expenditures, and assess the implication for ammunition savings.

Table 2.3 shows FY 2011 unit requirements, authorizations, and expenditures for small-arms, howitzer, and tank main-gun training ammunition.[10]

The data in Table 2.3 show that training ammunition requirements and authorizations greatly exceeded expenditures for all the weapon systems considered. Given that the Army does not procure training ammunition to support the strategies, but rather to replace actual expenditures, the current rate indicates that, all things being equal, even a substantial reduction in doctrinal requirements would not have led to real budget savings.

---

[9] The Longbow Crew Trainer (LCT) for the Apache was fielded during this period, but it was fielded as part of the fielding of the new attack helicopter, not to support an already fielded weapon system.

[10] The data in the tables in this section were provided by the DA Deputy Chief of Staff, G3/5/7. They came from the *Total Ammunition Management Information System* (TAMIS).

Table 2.3
Unit Training Ammunition Requirements, Authorizations, and Expenditures, FY 2011

	Rifle	Squad Automatic Weapon (SAW)	Machine Gun	155mm Howitzer	120mm Tank
Required	217,889,552	82,954,640	78,811,789	204,303	115,888
Authorized	191,173,423	73,078,123	65,619,480	196,147	82,450
Expended	58,408,636	20,970,222	23,486,757	112,142	13,061
Percentage of required expended	26.8%	25.3%	29.8%	54.9%	11.3%
Percentage of authorized expended	30.6%	28.7%	35.8%	57.2%	15.8%

Realizing that the recent low rates may have been affected by high levels of deployment, we also examined similar data for 2001, the year before the start of deployments to Afghanistan. These data, comparing the Army's total training ammunition requirements and expenditures for the same weapon systems, are shown in Table 2.4.

The data for 2001 show a much higher training ammunition expenditure rate relative to requirements. This rate is especially high given that authorized levels (not shown, but related to and generally lower than requirements) represent a ceiling and fully reaching them is not practically possible.[11] This is because a unit commander can only request up to authorized levels, and unforeseen circumstances almost always lead to using less than can be requisitioned.

This look at the 2001 training ammunition expenditure data indicates that units could feasibly come close to executing the DA Pamphlet 350-38 strategies in peacetime. The data also suggest that training ammunition usage per weapon system will likely increase as unit deployment rates decrease. Thus, given that the Army procures training ammunition to replace

Table 2.4
Army-Wide Training Ammunition Requirements and Expenditures, FY 2001

	Rifle	SAW	Machine Gun	155mm Howitzer	120mm Tank
Required	312,543,539	84,077,013	85,183,964	582,010	277,074
Expended	262,939,810	69,991,450	68,876,394	506,907	235,179
Percentage of required expended	84%	83%	81%	87%	85%

[11] The data in this table provide for only a general comparison with the 2011 data. The data in Table 2.3 are for units, while the data in Table 2.4 are Army-wide (e.g., they include TRADOC rifle training in Initial Entry Training). While a direct comparison of unit requirements and expenditures would have been desirable, it was not possible. The way data in the TAMIS are collected and aggregated changed over time as refinements and improvements were made. As a result, the exact analysis done for 2011 was not possible for 2001. Also, for 2001, we are only able to show total Army data, not authorized data. Comparing the Army-wide requirements and expenditures for 2001 with those for 2011 shows the same general trend as comparing the Army-wide requirements and expenditures for 2001 with the unit requirements and expenditures for 2011. A far greater percentage of requirements were executed in 2001.

expenditures, the actual cost of supporting DA Pamphlet 350-38 strategies could increase significantly in the near future.[12]

The Current State of Processes for Developing Weapon Training Strategies, Standards, and Support

In this section, we first outline the findings from our examination of the processes for developing weapon training strategies. We then present conclusions concerning areas where those processes could be improved.

Findings

The process for developing strategies has little centralized direction. We started our examination of weapon training strategy and standards development processes by looking for Army-level guidance on this subject and found little. General responsibilities are outlined in several publications, primarily Army Regulation (AR) 350-1, *Army Training and Leader Development*, DA Pamphlet 350-38, and AR 5-18, *Total Army Munitions Requirements Process and Prioritization System*. TRADOC has defined its overall processes for developing training standards and products in TRADOC Pamphlet 350-70-1, *Guide for Developing Collective Training Products*.[13]

But overall these publications do not offer specific guidelines for developing weapon training strategies or standards. This means that the weapon system proponents have the responsibility of developing these strategies and standards, but how their staffs do this is not standardized across proponents. This lack of standardization, coupled with the fact that weapon system training staffs have little understanding of current unit weapon system training practices, performance, needs, and technology use, greatly complicates Army-level assessment and resource allocation across weapon systems.

The overall process is evolutionary and is only staffed to support incremental change. To understand how these processes were implemented, we conducted extensive interviews with weapon training staffs at the Maneuver, Maneuver Support, Aviation, and Fires COEs.

Broadly speaking, proponents assume that their weapon training strategies and standards are sound and effective given the resource levels that can be reasonably allocated to weapon training, that the standards support combat requirements, and that the strategies provide proven paths for meeting those standards.

Proponents also contend that weapon proficiency is critical and that the strategies, and the resources that support them, should be protected. Moreover, they stated that the feedback they received from operational force commanders supported this view and that, in fact, much field feedback has indicated that higher—not lower—training ammunition authorizations are needed.

At the same time, they understand that strategies and standards must change to respond to new operational requirements, new weapon capabilities, and doctrine. Their major efforts have

---

[12] This would be offset by a smaller number of units as the Army's end strength is reduced.

[13] TRADOC used to have a detailed regulation outlining its systems approach to training and a series of supporting pamphlets. TRADOC Regulation 350-70, *The Systems Approach to Training Management, Processes, and Products,* and most of the supporting regulations and pamphlets have been rescinded without replacements. We were not able to review these earlier documents to determine if they outlined processes for developing weapon training strategies and standards.

been to examine how strategies should be adapted to these changes, to make other improvements, and to justify additional resources for needed adaptations. This means that the emphasis is on improving strategies to meet new requirements, not on making them more efficient.

While a more in-depth analytical process involving review and assessment would seem necessary for managing a program that is so important and involves such a significant amount of budget resources and unit time, training development resources are constrained across TRADOC. During our discussion with weapon training developers, they acknowledged that a larger research and analysis effort was needed but stated that their authorizations and staffing levels were not adequate for a larger effort.

The Army has a structured process for determining changes to weapon training strategies. While the development process is decentralized and evolutionary, there is a fairly structured process for approving changes. When a proponent determines the need for change, both the change and its rationale are reviewed by a DA-level Ammunition Working Group that meets semiannually and that is composed of representatives from DA, TRADOC proponents, and Army commands. If the working group accepts the change, it is forwarded to DA for action. If the change increases training ammunition requirements, it must compete against other Army needs. Getting added resources for any program is a difficult process, and full justification is always needed. Finally, if the change needed is large, review by DA's Training General Officer Steering Committee, and approval by the DA G3/5/7 is required. Even if approved by the DA G3/5/7, this priority must still compete for Army funding in the overall budget prioritization process.

The Army also has a structured process for determining unit training ammunition requirements and authorizations. The main source describing this process is AR 5-18, *Total Army Munitions Requirements Process and Prioritization System*. We reviewed this regulation and also discussed this process with key DA and U.S. Army Forces Command (FORSCOM) staff. Data on training ammunition were pulled from TAMIS, a web-based application for managing training ammunition requirements, authorizations, forecasts, requests, and expenditures.

From a unit perspective, the process for obtaining training ammunition starts with troop-owning commands determining their requirements (the quantity of training ammunition their units must have to execute their training programs) for the next year. Supported by an internal application in TAMIS, requirements are first computed by multiplying the weapon training ammunition requirements documented in DA Pamphlet 350-38 by the number of authorized weapons in the command's units.[14] Commanders can then adjust these base requirements as needed. For example, requirements could be increased to accommodate predeployment training needs not documented in DA Pamphlet 350-38, or decreased if deployments or other requirements would limit a unit's ability to execute the STRAC strategies.

In practice, commanders generally submit requirements that exceed the pure DA Pamphlet 350-38 strategy amounts recognized by TAMIS. When this occurs, the requests are reviewed and validated first by the chain of command, then by DA-level working groups, and then by the DA Deputy Chief of Staff G3/5/7. As a part of the review and validation process, previous training ammunition expenditure rates are considered. For example, if expenditure rates have been low, the validated levels may be reduced.

---

[14] Many units are not fully manned, and some soldiers may not be available for training. For example, attending a TRADOC school course. So the consideration that there are more weapons than soldiers who can be trained is one reason units may execute less than their training ammunition requirements.

A validated requirement does not necessarily translate into an authorization. The Army might not have enough ammunition of a particular type to meet validated requirements, and, as a result, authorizations may be reduced or other types of ammunition may be substituted.

Institutional understanding of analytical bases and history is limited. While the general belief is that the strategies and standards are sound and that the resources that support them are needed, there is little institutional knowledge of the analytical bases for many of these strategies, standards, and resource levels. For example, no one at the Maneuver COE was aware of any analysis that showed why 23 of 40 engagements is the required number of rifle hits for a soldier to be considered qualified or why qualification is required every six months. While there may well have been objective analytical underpinnings to these requirements at some point, these are no longer available to support the revision of strategies and standards or be used in defense of the resources needed to support these strategies.

Also, there may be a need for revision. As discussed in the previous section, while the strategies have undergone some changes over the last ten years, the overall structure and number of rounds supporting the strategies remain basically the same. Given that operational requirements and unit readiness processes have changed, whatever analytical underpinnings that exist may now be dated.

The exception we found was for tank gunnery. We were told that when the Maneuver COE recently revised its engagement standards for tank gunnery, they were based on computer battle-simulation results.

In addition to our inability to find analytical bases, our research indicates that it is likely that military judgment underlies many strategies and standards. In the case of one weapon system, the Avenger, a staff member recalled that the original standards and strategy were developed by a group of experts from different areas of the Army. This is not meant to say that military judgment–based strategies and standards are necessarily wrong. Military judgment should reasonably be a major component of the developmental process. However, a lack of institutional understanding of analytic bases and a large use of military judgment can complicate the revision and defense of a weapon training strategy.

The wide range of resources needed to support weapon training strategies come from many sources. Conducting weapon training requires ammunition, ranges, targetry, simulations, and other types of resources. As part of our review of weapon training–strategy processes, we found that these resources come from different Army Program Evaluation Groups (PEGs), programs, and program subcomponents. The Army manages resources under Management Decision Evaluation Packages (MDEPs).[15] Figure 2.1 shows the PEGs and MDEPs that support weapon training, illustrating the complexity of the resource relationships.

Weapon training resources come not only from the Training PEG MDEPs that provide simulators, ranges, and targetry, but also from those that provide resources for developing weapon system doctrinal products. Units also provide time and effort to plan, prepare, and conduct the weapon training.

---

[15] See AR 1-1, *Planning, Programming, Budgeting, and Execution System*, Washington, D.C., February 1994. The Army manages resources for its responsibilities under functional groups called Program Evaluation Groups (PEGs). There are six PEGs: Manning, Training, Organizing, Equipping, Sustaining, and Installations. Each PEG has multiple MDEPs. Individually, MDEPs describe a particular organization, program, or function and the resources associated with it. Collectively, MDEPs account for all Army resources.

Figure 2.1
PEGs and MDEPs That Support Weapon Training

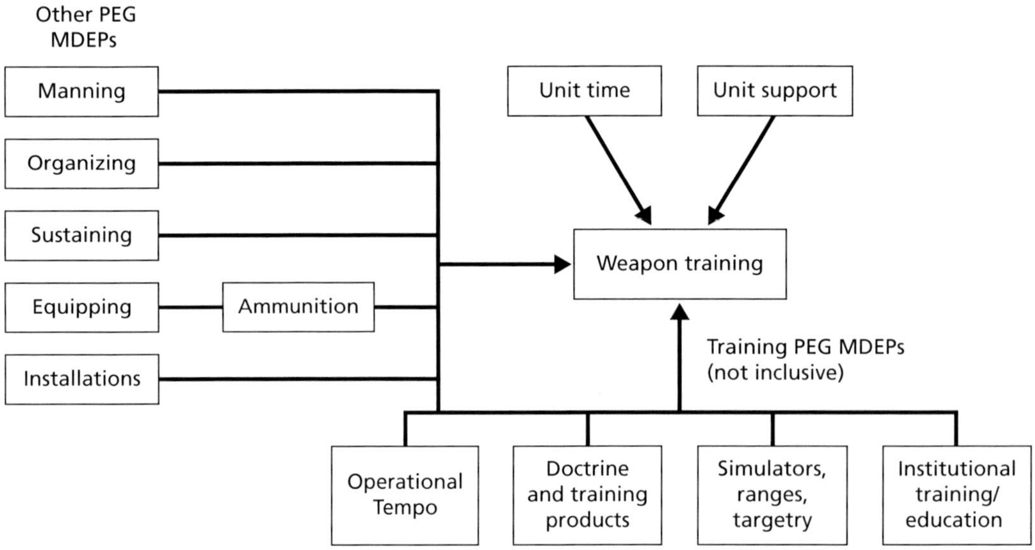

RAND *RR448-2.1*

But other resources come from other PEGs and their MDEPs, such as those that provide fuel and weapon repair parts (operational tempo) for unit training. The most important area where integration is needed is training ammunition, which is provided by the Equipping PEG. This arrangement especially complicates making changes the mix of live-fire and virtual modalities in weapon training strategies if an increase in simulators is needed to implement the change and up-front resourcing is needed before training ammunition savings can be achieved. The Vice Chief of Staff of the Army must approve such a shift. While this is not a major barrier, it does complicate the ability to make such changes.

There are other examples of resources provided by other PEGS. The level of unit manning and turnover is decided under the Manning PEG, and these affect the ability to conduct weapon training and the frequency of training needed to maintain crew and collective proficiency. To conduct training, the unit must have the weapon systems that it will deploy with from the Equipping PEG. Installation MDEPs provide much of the indirect support for ranges and simulators.

The point is that weapon training strategies require resources from many different sources. Resources are limited and demands are many. This complicates the requirement to coordinate for the resources needed to conduct weapon training.

Mechanisms for integrating weapon training strategies with other training strategy elements are informal. The higher the resource levels allocated to weapon training, the less time and resources will be available for competing training needs. To achieve the best possible readiness outcomes, training resources should be balanced across programs and requirements. However, overall unit training strategies (Combined Arms Training Strategies [CATS]) are developed by different staff elements at the proponents from those developing weapon training strategies, and we could find no formal mechanisms for ensuring these strategies are feasible and balanced when aggregated.

Likewise, and more directly related to the concern of improving the mix of virtual and live methods in weapon training strategies, there are no formal mechanisms for integrating live-fire and simulator capacities into comprehensive weapon training strategies or for calculating cost-benefit trade-offs between simulator and live-fire capabilities (specifically training ammunition and ranges).

From the proponent to the DA level, different staff organizations are responsible for developing weapon training strategies than for developing simulation requirements and fielding documents. In general, we found that weapon training staffs tend to focus on live-fire approaches, not on supporting the efforts of the Training Support System staffs to develop and field more effective weapon training simulator capabilities. This is not to say that such integration does not happen, but the systemic separation of the communities complicates integration.

There is little direct knowledge of unit programs, needs, or the use of technology. Weapon system proponents have no mechanisms for systematically gathering information on current unit weapon training practices. They generally know the overall level of training ammunition expenditures; how they vary by type; and that, in most cases, the levels expended are less than what they would be if all units executed all the activities outlined in the strategies. However, they have limited data on which activities are being done, how frequently they are being done, or which activities field commanders view as important. Even data on qualification rates or the frequency of qualification are not available to proponents, TRADOC, or DA staffs.

Similarly, while the overall training usage rates of simulators are collected and known to be low, there are no mechanisms for knowing which activities are being done or why usage rates are low. For example, are units using EST to sustain rifle marksmanship skills or meet qualification requirements such as the 10-meter machine gun tables? Are usage rates low because the simulation is seen as ineffective, because of access difficulty, or because of some other reason?

An understanding of operational units' needs and practices and the reasons behind these factors is important to improving weapon training strategies and their required support. For example, how frequently do soldiers or crews need to qualify to maintain needed proficiency levels? And what should be done to increase EST usage and its benefit on small-arms engagement skills? Moreover, an understanding of unit weapon training programs and their constraints is necessary for determining ammunition requirements and authorizations and adapting the weapon training strategies that underpin them.

Conclusions

Improved processes appear necessary to make the major changes needed to adapt to changed operational requirements and readiness processes. These findings should not be interpreted as stating that weapon training standards and strategies are wrong or that their evolutionary development processes are broken. Given tactical successes in Iraq and Afghanistan, it would not be reasonable to argue that the Army's weapon training standards and strategies are not successful, that they are not a key component of operational success, or that the evolutionary process of continual improvement does not support these results.

However, neither is it reasonable to assume that there is no need for improvement. As we will discuss later in this report, we could find no analysis of operational small-arms engagement proficiency in Iraq or Afghanistan being used to support the development and improvement of their training strategies. It is fair to say that the Army should review its weapon training strategies to ensure that they include current unit practices, can meet today's operational needs and threats, are feasible for units to execute, and are as efficient as reasonably possible.

Current processes do not include the data collection or analysis needed for such a review. Fundamental process changes would be needed to support such efforts.

Making such improvements will be a challenge. Limited staffing resources are authorized to support weapon training strategy development processes, and the actual staffing levels we saw generally fell below authorizations. They seem sufficient to support evolutionary change, but not to conduct the major research and analysis efforts (such as reexamining basic assumptions, revising standards to respond to changed threats, and analyzing improved mixes of live and virtual methods) that should be accomplished before making major changes to a successful and important program.[16] Moreover, the prospects for increased TRADOC staffing levels for any activity are not good.

---

[16] For example, when the Maneuver COE was tasked to examine alternatives for implementing improved training for non-stabilized mounted weapon systems, a large contracted support effort was requested and approved to supplement in-house analytic capabilities. Normal staff authorizations were not sufficient.

CHAPTER THREE

The Potential of Simulators to Improve Weapon Training Strategies and Make Them More Efficient

In this chapter we first outline our findings and then our conclusions on the potential of simulators.

Findings

There Is a Consensus that Simulators Can Improve Weapon Training and Make It More Efficient

Conceptually, weapon system simulators have considerable potential to augment live-fire training methods and improve weapon training effectiveness and efficiency. Potential improvements are outlined in Table 3.1.

In addition to increased training effectiveness, simulations offer the potential to reduce costs. Simulators can allow problem shooters and crews to be identified early and receive focused training. They can also support engagement skill sustainment. As a result, fewer soldiers/crews would need to refire to qualify and higher overall live-fire performance could be achieved. Finally, simulators have the potential to replace some live-fire training.

Table 3.1
Examples of Potential Simulator Improvements to Weapon Training

Potential Improvement	Examples of Why Improvement Possible
Increase frequency of weapon training	Simulator does not require drawing, transporting, and turning in ammunition
	No range and ammunition availability issues
	Weather does not affect ability to use simulator
Increase weapon-performance feedback	More detail (round impact, point of aim, weapon sight tracking, etc.) than possible in live-fire
	Automated diagnostics possible
	Better scoring of area-fire engagements
Train tasks under conditions that are not practically possible on live-fire ranges	Engagement of moving targets
	More realistic targets
	Wide sectors of fire and larger engagement areas
	Few safety limitations (e.g., engagements in close proximity to maneuvering soldiers possible)

Direct Research on the Capabilities of Gunnery Training Simulators Is Limited

While simulators have been used extensively for many types of training, including weapon training, research that definitively proves their ability to train weapon engagement skills is limited. As discussed in Chapter Two, we found few studies directly showing that simulator proficiency was related to live-fire proficiency.[1] In many cases, the Army seems to have reasonably assumed that if a simulator realistically replicated the act of engaging targets with service ammunition using the actual weapon system, then simulator training would effectively train live-fire engagement skills.

However, not testing training transfer from simulator training to actual weapon system engagements can prove a major mistake if for some reason the simulator does not realistically replicate important engagement procedures and techniques. As an example, the Army's training simulator for a now-obsolete anti-tank guided missile, the Dragon, had such an issue. The simulator scored gunner performance based on the time the sight crosshairs remained on the target. In an actual engagement, if the crosshairs were off the target, it would be necessary to ease them back to the target. Thus, in a simulator engagement, the gunner would get credit for hitting the target using engagement techniques that could result in a live-fire miss. The net result was that the simulator taught improper engagement techniques and could have had negative effects on gunner performance.[2]

Studies of Rifle Marksmanship and Tank/BFV Gunnery Support the Benefit of Simulators for Weapon System Training

There are two areas where fairly extensive, definitive research on the capabilities of simulators to support weapon training has been conducted—tank/BFV gunnery and rifle marksmanship. This research is described below.

Tank/BFV Gunnery. A large number of studies were conducted in the late 1980s and 1990s examining tank and BFV gunnery training simulations and devices.[3] In this section, we summarize the key findings of those studies.

The studies examined a broad range of tank and BFV devices, including simulators, desktop-type trainers, and laser and sub-caliber devices. The capabilities of each training tech-

---

[1] For example, see Patrick E. Corcoran, *Simulation and Analysis of the Training Effectiveness Analysis-TOW (TEA-TOW) Flight Data*, U.S. Army Materiel Systems Analysis Activity (AMSAA), April 1980. This study found that the flight data of the simulation were accurate but did not find that training on the simulator resulted in live-fire proficiency. Likewise, discussions of the effectiveness of air-defense missile simulators with training staff at the Fires COE indicated that direct testing was limited. Such tests, especially those involving expensive weapons and realistic targetry, are difficult and expensive, which is an increasingly difficult issue as budget pressures impact on the Army's ability to do such testing.

[2] One of the authors was a member of the Infantry School's combat development staff for the Dragon during this period.

[3] This section summarizes the principal findings from the following research documents: Ronald E. Kramer and Wade C. Rowatt, *A Review and Annotated Bibliography of Armor Gunnery Training Device Effectiveness Literature*, Alexandria, Va.: U.S. Army Research Institute, ARI Research Report 1652, November 1993; Bruce S. Sterling, *Relationship Between Platoon Gunnery Training and Live Fire Performance*, Alexandria, Va.: U.S. Army Research Institute, ARI Research Report 1701, September 1996; Charles R. Hughes, Walter B. Butler, Bruce S. Sterling, and Arvid W. Berglund, *M-1 Unit Conduct of Fire Trainer-Post Fielding Training Effectiveness Analysis*, Alexandria, Va.: U.S. Army Research Institute, TRAC-WSMR TEA-16-87, ACN 67571, April 1987; Joseph D. Hagman and John E. Morrison, *A Strategy for Effective Device-Based Tank Gunnery Training in the Army National Guard*, Alexandria, Va.: U.S. Army Research Institute, ARI Research Report 1694, June 1996; and Charles R. Hughes, Maryann Morales-Steigely, and Max L. Musser, *M2/M3 Unit Conduct of Fire Trainer; Post-Fielding Training and Effectiveness Analysis*, Alexandria, Va.: U.S. Army Research Institute, TRAC WSMR TEA-90-016 CAN 023378, August 1990.

nology in terms of accurate replication of actual weapon system operation varied. Here we focus on findings regarding the COFT for the tank and BFV because these are the Army's main tank/BFV gunnery training simulators.[4]

In the COFT, the vehicle commander and gunner occupy simulated vehicle compartment crew stations that provide stimulus of, and response to, crew actions in a manner closely approximating the actual system. The visual subsystem simulates tactical scenes consisting of a variety of terrain, man-made cultural features, and potential target types so that the vehicle commander and gunner can be trained in target detection, classification, and gunnery under simulated combat conditions. COFTs have an instructional subsystem (matrix) that provides progressive training exercises in which crew interact with each other in response to visual and aural cues while performing a variety of special purpose task training and tactical firing exercises. A training management program provides the data necessary to assess crew gunnery proficiency and direct crew training.[5]

The studies found evidence that the COFT and most of the other devices provided training benefit and improved skills. The tank COFT Post Fielding Training Effectiveness Analysis (TEA) showed that higher matrix achievement and more COFT training resulted in statistically significant increases in live-fire hit probabilities and faster opening-fire times. While these correlations were significant, they were not strong enough to precisely predict qualification performance. The researchers concluded that the likely reason for this lack of a precisely predictive relationship was that other components of tank gunnery training also influence gunnery performance. Another TEA conclusion was that the COFT could make a substantial contribution to tank crew sustainment if used correctly.

An interesting finding was that the COFT hit percentages for stationary targets were higher than for live fire, but the converse was true for moving targets. Soldier comments about differences between the sensitivity of gunner controls and stabilization features between the tank and COFT suggested that a lack of simulator fidelity was the source of this difference. This, along with the Dragon simulator results, suggests that achieving simulator fidelity is important. Of course, this also adds to simulator cost and development time.

None of the studies looked at using training devices or simulators to replace live-fire events. Instead, they examined their use to supplement and enhance live-fire programs. However, they did support the conclusion that gunnery simulators could reduce the amount of training ammunition needed to reach similar qualification results and that the time required for live-fire training could be reduced.

Rifle Marksmanship. A fairly extensive set of studies on rifle training devices and simulators also exists.[6] As with the tank and BFV studies, these covered a wide range of devices

---

4 There are a number of different versions of Tank and BFV COFTs. The most advanced versions for the M1A1/2 tanks and Stryker Mobile Gun Systems (MGSs) are called the Advanced Gunnery Training System (AGTS). For simplicity, we use the generic term *COFT.*

5 From DA Pamphlet 350-9, *Index and Description of Army Training Devices*, Washington, D.C., May 2010.

6 The small-arms simulations studies we reviewed were the following: Kenneth L. Evans, Jean L. Dyer, and Joseph Hagman, *Shooting Straight—20 Years of Rifle Marksmanship Research*, Alexandria, Va.: U.S. Army Research Institute, ARI Special Report 44, October 2000; Kevin Murphy, James L. Farr, and Greg Loviscky, *Study to Quantify the Benefits and Costs of Simulated Versus Live-Fire Training at USMC Ranges*, University Park, Penn.: Penn State University, January 2007; Robert A. Wisher, Mark A. Sabol, and John A. Ellis, *Staying Sharp: Retention of Military Knowledge and Skills*, Alexandria, Va.: U.S. Army Research Institute, ARI Special Report 39, May 1999; Joseph D. Hagman, "Using the Engagement Skills Trainer to Predict Rifle Marksmanship Performance," *Military Psychology*, No. 10, Vol. 4, 1998, pp. 215–224; Charles Hughes

and simulators, but their major focus was on the EST and the Laser Marksmanship Training System (LMTS). Both training systems provide realistic target engagement with automated feedback for a range of small arms.

The EST is the Army's primary small-arms marksmanship simulation trainer.[7] The normal system has ten firing lanes and the basis of issue is one per brigade combat team or its equivalent. The EST is a computer-operated simulator that provides the shooter with a realistic opportunity to engage visual targets with simulated weapons that physically replicate (noise and recoil) shooting actual weapons. The strike of rounds on targets is ballistically accurate, and the software provides for feedback (e.g., where the target was hit and how the sight picture changed). The simulator capabilities include an eye-safe laser; sensors to measure trigger pressure, cant, and ammunition magazine/belt status; and a compressed air operating system to provide recoil. "Shoot, Don't Shoot"; Collective; and Marksmanship scenarios are also preloaded onto instructor/operator stations.

The LMTS is an indoor/outdoor rifle marksmanship–training device that simulates weapon training events to prepare individuals for live-fire qualification.[8] The LMTS uses a laser transmitter on the individual soldier's issued weapon to engage a reflective target. A scoring device shows where an actual shot would have hit a target at various ranges. Since it is light, transportable, has a self-sustained power source, and requires no fixed facilities support, it can also be used at field sites. The LMTS provides real-time quantitative feedback on key fundamentals of marksmanship, including steady position, sight alignment and picture, breath control, and trigger squeeze.

Another training simulator that was examined was the Multipurpose Arcade Combat Simulator (MACS), an inexpensive rifle marksmanship trainer that was developed to overcome the training problems encountered because of insufficient facilities. MACS was a computer-based, part-task weapon trainer. The system components included a weapon (e.g., a demilitarized M16 rifle) with an optical focus light pen attached to the barrel, a computer, and software appropriate to the characteristics of the weapon. The system allowed shooters to practice basic skills by firing at targets at scaled ranges displayed on a computer screen with diagnostic feedback provided. The system was validated as a teaching device.

The studies found that all of the small-arms trainers provided some level of skill acquisition, especially for soldiers with poor shooting skills.

The EST and LTMS were the most studied devices, scores on both simulators were found to be predictive of live-fire qualification scores, and both were judged as having the potential to sustain live-fire skills. In the area of skill sustainment, the EST TEA examined substituting practice on EST for live-fire practice in preparation for live-fire qualification and found no significant difference between groups of soldiers who practiced using live fire and those who practiced on EST. However, the findings were similar for a control group that did neither; thus the benefits of EST were not fully established.

and Kathy L. Nau, *Engagement Skills Trainer (EST) 2000, Training Effectiveness Analysis (TEA)*, Fort Monroe, Va.: TRADOC Analysis Center, March 2008; and Claude Miller and Michelle Zeisset, *Engagement Skills Trainer (EST) 2000, Post Fielding Training Effectiveness Analysis (PFTEA)*, Fort Monroe, Va.: TRADOC, WSMR-PFTEA-02-025, September 2002.

7 For further information on the EST, see DA Pamphlet 350-9, *Index and Description of Army Training Devices*, May 2010.

8 For further information on LMTS, see DA Pamphlet 350-9, *Index and Description of Army Training Devices*, May 2010.

Although these studies could support the conclusion that rifle marksmanship simulators could develop marksmanship skills and reduce the number of rounds needed to build and sustain those skills, they also showed the need for live-fire training, especially for physical weapon-operation activities such as weapon loading and immediate action drills for misfires.

Many Obstacles and Limitations Hamper the Expanded Use of Simulators and Other Technologies in Weapon Training Strategies

While simulators have many potential benefits in terms of supporting more efficient and more effective weapon system training and some studies provide evidence of this potential, many factors must be considered before concluding that a major shift in strategies to save training ammunition costs is possible or desirable. The remainder of this section details the most important of these factors.

Strategies already direct the extensive use of available simulator capabilities. As discussed in Chapter Two, the extensive use of simulators is currently included in most weapon training strategies, especially those for weapon systems with high-cost ammunition. For example, anti-tank guided missiles and air defense systems have a simulator-based training strategy with little or no live fire.

Moreover, it is not clear that the number of fielded simulators is sufficient to expand their use in current strategies. As a key example, current rifle and pistol qualification standards require CBRN and unassisted night-fire qualification twice a year in EST, as well as practice for this qualification. Additionally, pistol and rifle strategies call for the use of EST in practice quarterly. Likewise, 10-meter qualification and EST practice for automatic rifles and 7.62-mm machine guns are an EST requirement.

As mentioned above, each brigade or brigade equivalent (approximately 3,500 soldiers) has one ten-station EST. We made some simple calculations to determine what this fielding generally means in terms of the ability to support current small-arms training strategies.

Making reasonable, but somewhat optimistic, assumptions that a brigade's EST is available for training 240 days per year, and that it takes a group of soldiers about two hours to go through a training period, for a total of four training periods per day, we calculated a capacity of about 2.7 sessions for each soldier in a Brigade Combat Team (BCT) or BCT equivalent per year (240 days × 4 periods per day × 10 stations/3,500 soldiers in a BCT). With effective scheduling of soldier training and EST maintenance, this is enough capability to support rifle, pistol, and machine gun qualification requirements, but not Preliminary Marksmanship Instruction (PMI) or other weapon training strategy requirements (e.g., unstabilized mounted weapon systems; "Shoot, Don't Shoot"; and AT4 anti-tank weapon training). Thus, the use of EST in rifle, pistol, automatic rifle, and machine gun strategies exceeds EST's reasonable throughput capability. Moreover, these throughput assumptions are optimistic. For example, training 40 soldiers per day would require very effective scheduling and training execution.

Units do not seem to be executing current strategies; so changing them will not necessarily result in real savings. As we saw in Chapter Two, training ammunition expenditure rates and simulator usage indicate that units do not follow current strategies. Given this consideration, there is no basis to support the conclusion that better use of simulators would save resources. Units may be making reasonable use of simulators now, and whether the increased use of currently fielded simulators would improve actual strategies and outcomes cannot be demonstrated without further examination.

Expanding simulator capacity will not result in near-term savings because additional time and investment would be required. Fielding new simulators or expanding currently fielded simulator capability during a time when the Army is looking for savings would be difficult because it would require an initial investment of resources, and these resources would have to be spent before any training ammunition savings could be realized.[9] Furthermore, there would be long lead times before new capabilities could be fielded because it takes many years to develop, test, and field new simulators. Five years was the most optimistic estimate we were given by weapon-proponent staffs. Also, generating the support to develop a new simulator capability takes effort. In the last ten years, the only new simulator among the weapon systems we examined was the Longbow Crew Trainer, a flight and gunnery simulator for the latest version of the Apache attack helicopter. Moreover, the LCT was fielded as part of the fielding of the new helicopter, not as an added training capability for a fielded weapon system.[10]

Even sustaining the capabilities of fielded simulators can be a difficult process. Implementing upgrades to keep the EST up-to-date with the current operating environment (e.g., new scenarios) and force modernization requirements (e.g., new sights) took considerable time. The upgrades were approved in 2007, after several years of effort to develop the requirements document. The fielding process was not completed until 2012.[11]

Operational force leaders would likely resist changes that would result in less live-fire capabilities. A final consideration is that there are strong indications that Army leaders highly value live-fire training and would voice strong opposition to reduced training ammunition allocations in favor of increased simulator use. This concern was repeatedly brought up during our visits to the COEs. The high value the field force places on live-fire training was reinforced by a recent unpublished RAND study of unit commanders and staff, which showed that they thought live-fire ranges and ammunition were the most beneficial types of training enablers.[12]

An Examination of the Fire Support Combined Arms Tactical Trainer (FSCATT) Illustrates Some Key Issues with Leveraging the Potential of Simulators

An illustrative example of the difficulty and limitations of trying to move to a more simulator-based approach to weapon system training is the Army's efforts to develop the FSCATT for the Paladin field artillery system.[13] The concept for the simulator was initially developed in the early 1990s, when the field artillery community was told it could expect to face major reductions in training ammunition allocations. At a cost of $600 or more per round, depending on

---

[9] For example, an EST with 10 stations costs about $500,000, and a 5.56mm round costs about $.36. This means that 1.4 million fewer rounds would have to be fired to buy back an EST. Saving this amount would require that units actually fire fewer rounds, not just increase EST use.

[10] The LCT was fielded by the Apache Program Manager's funds as part of the system and part of the Army's equipping PEG. Funding and sustainment of the simulators for already fielded weapon systems come from the Army's Training PEG.

[11] See Department of the Army Memorandum, *Approval of the Engagement Skills Trainer, Increment II (EST II) Capability Production Document (CPD),* Washington, D.C., February 2006.

[12] This survey was conducted in 2010. 133 unit commanders and staff members were asked to rank order the benefit of 17 different types of live, virtual, or constructive training enablers. Five of the enablers directly supported live-fire training. The top four were live-fire enablers, and a fifth live-fire enabler was ranked seventh.

[13] After reviewing U.S. Army Training Support Center, *The Operational Requirements Document for the Field Artillery Fire Support Combined Arms Tactical Trainer (FSCATT),* memorandum, March 1993, we interviewed training developers at the Fires COE to determine the history and current capabilities of the FSCATT.

the type of ammunition, live-fire practice is expensive, so developing a simulator capability can make economic sense. After the initial fielding investment, multiple units can train on a large number of engagements on a single simulator set for far less cost and effort than would be required for added live-fire training.

Such a simulator could considerably enhance field artillery training. Even with about 160 rounds authorized per howitzer annually, that allocation does not provide enough ammunition to practice all the important types of fire missions that are needed for full field artillery unit proficiency. Moreover, it is difficult to assess the accuracy of the three-component field artillery system (howitzer crews, Fire Direction Center personnel, forward observers) in a live-fire exercise. When a round misses the target, each component of the system could be a source of the error. Precise diagnosis is far easier for fire missions executed in a simulator, where the elements of target location, call for fire, gunnery computation, and crew performance can all be retrieved and compared with "should hit" data.

The concept of the FSCATT, as outlined in the requirement document, was for a "closed loop" system, or a simulator that could exercise the three components of the field artillery gunnery system working either together or separately.[14]

The original requirement document called for a battery set of self-propelled howitzer crew simulators (so that the six howitzer crews in a battery could be exercised at the same time) and outlined a requirement for a "strap-on" system in which FSCATT system components could be attached to either towed howitzers or to Paladin systems for the Army National Guard (ARNG).[15] It also specified a capacity to link to maneuver unit Close Combat Tactical Trainer (CCTT) simulator/simulation sets so that field artillery units could train with maneuver units.[16] The concept was to have one set at every major Army AC installation and strap-on sets for towed field artillery units and at selected ARNG armories.[17]

However, the development and fielding process of the FSCATT never produced a system that fully achieved this concept. The strap-on system was never developed, thus the current FSCATT cannot support towed artillery training or training at ARNG armories. Also, given the level of funding approved, the decision was made to field sets with a single crew simulator rather than a battery set, basically eliminating the capability to train howitzer batteries as part of a closed-loop system or to save significant amounts of training ammunition. Likewise, the capability to integrate with CCTT was never developed. Finally, when the Army developed a separate Call for Fire (CFF) training simulator for forward observers and an AFATDS training system for Fire Direction Center personnel, these were not integrated into FSCATT, so there is duplication of training systems in these areas.

Even before its development had been completed, FSCATT was becoming obsolete. Updated versions of the AFATDS were being fielded, and the simulator's system was not updated. Likewise, new types of artillery munitions were fielded, and these could not be

---

[14] This description is based on U.S. Army Training Support Center, 1993.

[15] ARNG training of this type normally takes place on weekends at local armories, typically with one battery at each armory. Procuring enough expensive simulators to have one at each training site would not be feasible. But these armories typically have a Paladin at each site, so the lower-cost strap-on versions could meet ARNG needs at a more reasonable cost.

[16] The CCTT is a training system that has crew-manned tank and BFV simulators and has work stations that allow the play of other maneuver battalion combat, command and control, and support elements in a virtual combat exercise.

[17] ARNG units train on weekends in dispersed armories, and only the strap-on version could reasonably support weekend training.

trained using the FSCATT system. Also, a new version of the Paladin vehicle, different from the one in FSCATT, was being fielded.

The field artillery training and material development communities made major efforts for over five years to obtain funding to upgrade FSCATT and deal with these issues, but only recently has the FSCATT system been updated to align with current equipment and munitions and integrate with the other field artillery training systems.

This examination of the FSCATT shows that the concept of using new simulator technology to reduce training ammunition requirements and enhance weapon training can be very hard to achieve. Moreover, it underscores a theme we heard from all the COEs: Translating a seemingly reasonable training capability into an actual capability is challenging. An objective cost-benefit analysis is needed to ensure that the development and fielding of the training capability will be feasible and supported. Moreover, given current pressures to reduce budgets and the issues encountered in expanding or even maintaining current simulator capabilities, obtaining near-term funds to create possible long-term savings will be a challenge.

Also, it is necessary to prove that there will be training benefit. A strong case must be made in argument for a system that many may not believe will train soldiers as well as traditional live-fire methods. It was not clear that the field artillery training community would willingly surrender any significant portion of its training ammunition allocations to fund simulators without strong proof of their training benefit, and to date, no TEA has been done to demonstrate the benefit of the FSCATT. In the case of the FSCATT, while current efforts to improve the fielded capability may make sense, from a larger perspective, it is not clear that the current and future benefits of the fielded training system have been worth the total cost and effort expended to date.

Conclusions

This examination of the potential of simulators leads to the following conclusions:

- While simulators have the potential to improve weapon training effectiveness, whether the improvements are achievable or worth the their cost are questions that must be examined before deciding to develop and field added simulator capabilities.
- A TEA should be done early in development to ensure that dollars and effort are spent to develop a system that can yield real training benefit at a reasonable cost. Likewise, training transfer should be proven before procurement. Fielding technologies that do not train effectively or are not utilized wastes resources.
- The potential for increased simulator use to significantly lower training ammunition costs remains uncertain. Simulators already play a major role in training strategies, and the systems with high-cost munitions already make extensive use of simulators. Moreover, it is not necessarily true that increased simulator fielding or use would reduce actual unit training ammunition expenditures while maintaining proficiency.

CHAPTER FOUR

Current Small-Arms Training Strategies

In Chapters Two and Three, we examined the Army's process for developing weapon system training strategies and the potential of simulators to improve weapon training and make it more efficient. We did this by examining a wide set of weapon systems. In this chapter, we begin our examination of directions the Army could take to improve these strategies by making them more relevant to current and near-term operational requirements, improving their support to ARFORGEN unit readiness processes, and making them more efficient.

We do this by examining three small arms: the M16/M4 rifle, the M249 SAW, and the M240B 7.62mm ground-mounted machine gun. In addition to being a small enough set of weapon systems to examine in detail, we chose to look at small arms for two other reasons:

- All soldiers who operate in combat use small arms; thus they are important to operational success and force protection.
- Because there are so many of these weapon systems across all Army units, even though the per-round costs are low, small-arms training ammunition represents a large portion of the Army's annual ammunition procurement budget.

We examined these weapon systems' training strategies and standards in detail to determine the degree to which they reflect the engagement skills needed for combat success and support achieving these standards in the context of ARFORGEN readiness processes. Here, we present the key findings arising from this comparison and draw conclusions about areas where these strategies and their standards could be improved. As part of this effort, we also examined the small-arms training strategies of the U.S. Marine Corps (USMC) and several other nations and compared them to the Army's strategies.

Small-Arms Training Strategies

In this section, we examine and analyze small-arms training strategies in detail.

Rifle, Machine Gun, and SAW Training Strategies and Standards

Key elements of M16/M4 rifle, M249 SAW, and M249 machine gun training strategies and qualification standards for AC soldiers, crews, and units are summarized in Table 4.1. The

Table 4.1
AC Small-Arms Training Strategies

	M16/M4 Rifle	M240B Machine Gun	M249 SAW
Individual/crew qualification events/ standards/frequency (rounds)	Zero/SA (18) each sight Unassisted Night/7 of 30 hits/ SA (0-in EST) CBRN/11 of 20 hits/SA (no rounds-done in EST) Record fire/23 of 40 hits/SA (40) Done with each primary sight	10-meter fire/63 of 91 hits/ SA (no rounds-done in EST) Zero/SA(28) Transition Fire/7 of 11 hits/ SA (154) Zero TWS /SA 36 Transition fire TWS/6 of 11 hits/SA (154)	10-meter fire/35 of 51 hits/SA (no rounds-done in EST) Zero/SA (12) Transition Fire/7 of 11 hits/SA (66) Done with each primary sight Zero TWS/SA (36) Transition Fire TWS/no standards listed /SA (54)
Additional individual/crew event/frequency (rounds)	PMI/Q (0-EST supported) Practice record fire (40) ARM/SA(200)—Infantry only	PMI/Q (0-EST supported) Practice Transition Fire (154) Practice TWS/S (308) Zero iron sight/A (28) Transition record and practice iron sight/A (308)	PMI/Q (0-EST supported) Zero iron sight/SA (12) Practice Transition Fire/Q (144) Iron sight practice and Transition/A (132) Practice TWS/SA (96)
Infantry collective qualification events/ frequency (rounds)	Squad/Plt LFX/Q (120)—No engagement standards Plt/Co LFX/A (120)—No engagement standards	Squad/Plt LFX/Q (275)—No engagement standards Plt/Co LFX/A (275)—No engagement standards	Squad/Plt LFX/Q (275)—No engagement standards Plt/Co LFX/A (275)—No engagement standards
Non-infantry collective qualification events/ frequency (rounds)	Squad/Plt LFX/A (120)—No engagement standards	Squad/Plt LFX/A—No engagement standards (275)	Squad/Plt LFX/A (275)—No engagement standards
Additional Infantry Collective events frequency (rounds)	Fire and Move LFX/Q (20) CALFEX/A (120)	CALFEX/A (200)	Fire and Move LFX/Q (70) CALFEX/A (200)
Ammunition required for strategy individual/crew, collective (total number of rounds)	Individual • Infantry with CCO (602 [includes ARM]) • Non-infantry with CCO (502) Collective • Infantry (1,010) • Non-Infantry (120) Total • Infantry with CCO (1,612) • Non-infantry with CCO (622)	Crew (2,004 with MGO and TWS) Collective • Infantry (1850) • Non-infantry (275) Total • Infantry with MGO and TWS (3,854) • Non-infantry with MGO and TWS (2,279)	Individual (804 with MGO and TWS) Collective • Infantry (2,130) • Non-infantry (275) Total • Infantry with MGO and TWS (2,934) • Non-infantry with MGO and TWS (1,079)

NOTES: A = annual; ARM = advanced rifle marksmanship; CALFEX = combined arms live-fire exercise; CBRN = chemical, biological, radioactive, nuclear; LFX = live-fire exercise; CCO = Close Combat Optic; MGO = Machine Gun Optic; PMI = Preliminary Marksmanship Instruction; Q = quarterly; SA = semiannual; SAW = Squad Automatic Weapon; TWS = thermal weapon sight. The number in parenthesis is the number of rounds per event.

table outlines weapon training events, including the ammunition required for the events and the number of hits required for individual/crew qualification.[1]

The data in this section were drawn from DA Pamphlet 350-38, which outlines the weapon training events required for qualification, other events in the strategy, the frequency of the events, and the number of rounds authorized for each event. FM 3-22.9, *Rifle Marksmanship M16/M4 Series Weapons*, and FM 3-22.68, *Crew-Served Machine Guns 5.56-mm and 7.62-mm*, were also examined to provide detail on the standards for each event.

In the next two sections, we further examine individual/crew and collective strategies and standards.

Individual/Crew Small-Arms Training Strategies and Standards

Examination of Table 4.1 reveals many similarities across strategies. All call for quarterly preliminary marksmanship instruction.[2] Individual and crew qualification are required semiannually. Most of the rounds in the individual/crew strategies are included in the semiannual zero, qualification, and practice qualification events.

But there are also differences in these strategies. An individual armed with a rifle is required to qualify on each primary sight; that is, if the soldier is issued a Close Combat Optic (CCO) and a Thermal Weapon Sight (TWS), the soldier must qualify twice every six months, once with each sight on what is basically the same qualification course. Rifleman equipped with a CCO must also qualify with the backup iron sight annually. For the machine gun and SAW, backup iron sight qualification is not required, but it is an annual event supported by the strategy.

The individual/crew qualification standards for these weapons also have many similarities in that they involve a series of live-fire engagements of stationary targets and groups of targets at various ranges. There are time limits for each engagement, and the number of target hits determines individual/crew qualification.

EST plays a role in the strategies and qualification standards for all three weapons.[3] It supports PMI for all the weapons. However, the use of EST for qualification differs for each weapon. Machine gun crews and SAW gunners have an EST qualification event—the 10-meter table.[4] This is a short-range event in which a series of targets are engaged in sequence, exercising the gunner's ability to engage multiple targets in succession. Unassisted night fire and CBRN are the two rifle qualification events (tables) conducted on the EST.

There are some differences among these weapon systems. Rifle engagements are semiautomatic, and machine gun and SAW engagements are automatic, short-burst engagements. CBRN engagements also differ. For the rifle, there is a CBRN qualification table that is done on the EST. SAW qualification includes one live-fire engagement while wearing a protective mask, and there are no CBRN engagement requirements for the machine gun. Both the machine gun and SAW have a table that is fired with night vision goggles. This table is similar

1 DA Pamphlet 350-38 also authorizes blank ammunition for tactical training events, but these are not included in this section. As will be discussed later, collective live-fire exercises above crew level do not have proscribed engagement standards in terms of targets to be engaged and percentage of hits.

2 How this instruction is conducted is outlined in the weapon system's FM.

3 These qualification requirements can also be done live fire, but the strategy calls for EST use if available. The EST is available for almost all AC soldiers assigned to units.

4 In weapon training strategies, many structured training events are called tables.

to the daylight table but with somewhat reduced ranges. The night-fire qualification requirement for the rifle is a table fired using the rifle's night sight, but there is no night vision goggle qualification requirement for the rifle, nor does DA Pamphlet 350-38 authorize ammunition for this type of training.

While the basic individual/crew strategies for each weapon are the same for all soldiers, infantry soldiers are authorized to conduct additional rifle training events in that they are resourced to conduct Advanced Rifle Marksmanship (ARM) semiannually.[5] ARM includes engagements from advanced firing positions, rapid semiautomatic fire, burst fire, moving shooter fire engagements, moving target engagements, short-range/reflex-fire engagements, and Squad Designated Marksman training. While there is a range of combat-relevant events in ARM, only a few can be conducted with the 200 rounds authorized per year.

Another difference is that DA Pamphlet 350-38 provides for enhanced rifle and SAW training and qualification for infantry soldiers assigned to Stryker Brigade Combat Teams (BCTs) that are not shown in Table 4.1. Squad Designated Marksmen (SDM) in Stryker units are authorized an additional 196 rounds per year to conduct the recommended SDM training program. And all their riflemen are authorized an additional 780 annual rounds for short-range marksmanship and are required to qualify on this skill quarterly. Likewise, Stryker unit SAW gunners are authorized an additional 800 rounds for short-range marksmanship and required to qualify quarterly. According to a Maneuver COE weapon training staff member, these differences are not because the Stryker infantrymen have a greater requirement than other infantrymen, rather because the program, which was approved when the Stryker BCT was an Army priority, while desirable, was not seen as affordable for all infantry units.

Collective Small-Arms Training Strategies and Standards

Except for ARM and Stryker infantrymen, all soldiers, regardless of branch or type of unit, have the same individual and crew strategies and qualification standards for each type of small arm.

However, this is not true for collective live-fire training. Infantry units are required to conduct a far larger number of collective events for qualification and have additional authorized events. The infantry qualification collective events in the strategy are quarterly squad/platoon LFXs and semiannual platoon/company LFXs. Additionally, infantry units have a CALFEX authorized annually and the rifle and SAW strategies have a quarterly Fire and Move live-fire event.

The result of this large number of collective events is that the number of rounds authorized to support infantry collective live-fire strategies is larger than the amount supporting individual training for the rifle (1,010 compared with 602) and SAW (2,130 compared with 804), and about the same for the machine gun (2,004 compared with 1,850). This means that it is important to examine collective live-fire events when considering options for improved small-arms training efficiency.

While a large amount of the DA Pamphlet 350-38's annual training ammunition authorizations are allocated to collective LFXs, there is no direct requirement to assess engagement skills in these events. In its guidance on LFX evaluation standards, Training Circular (TC)

[5] This is because ARM is listed in DA Pamphlet 350-38 as an infantry collective event, even though the ARM events are actually for individual soldiers.

7-9, *Infantry Live Fire Training*, lists collective tactical maneuver and leader collective tasks, but no engagement tasks or standards (such as types of engagements, ranges, and hit percentages).

Differences Between Doctrine and DA Pamphlet 350-38 Strategies and Standards

There are two areas of difference between the doctrine in the FMs and the strategies and standards in DA Pamphlet 350-38: the number of events and the types of standards. These are discussed below.

Number of events. The first issue is that the strategies and standards in rifle and machine gun/SAW FMs call for more live-fire training events than are authorized in DA Pamphlet 350-38. An important difference is that the machine gun FM states a requirement for the assistant machine gunner (as well as the machine gunner) to qualify, but this requirement is not supported in DA Pamphlet 350-38.

The rifle marksmanship FM includes live-fire short-range grouping exercises and known distance (KD) shooting to develop basic shooting skills as elements in the rifle training strategy, but DA Pamphlet 350-38 does not recognize these events.[6] The issue is that there are limitations to improving basic marksmanship skills on the record-fire events. While record-fire practice and qualification on pop-up targets provides a reasonable test of shooting skills, they do not yield what the FM calls "downrange feedback" (i.e., knowing more precisely where on the target a round hits). The shooter and coach know only if the target was hit, because the target falls down when hit. As emphasized in the FM and reinforced in our discussions with rifle marksmanship experts, downrange feedback is important to diagnosing, correcting, and improving basic rifle-shooting skills. A shooter must have good basic skills to fire a tight shot group (i.e., a series of rounds fired at the same aim point that strike close to each other). A tight shot group is key to getting a good zero,[7] and shooting well requires a good zero. Thus, grouping exercises can be an important event for improving marksmanship skills.

Also, the rifle FM, as discussed above, outlines additional ARM training events, but DA Pamphlet 350-38 does not authorize enough ammunition to fire all of these events for infantry units. For soldiers other than infantry, no ammunition is authorized for ARM. The machine gun/SAW FM also has advanced crew gunnery tables, but DA Pamphlet 350-38 does not authorize any training ammunition for these tables.

As shown in Table 4.1, most of the individual/crew annual training ammunition authorizations are in record-fire qualification and practice. While commanders are not required to implement all the elements of the DA Pamphlet 350-38 strategies, Army policy is that they adhere to its qualification requirements. To the degree that commanders follow this policy, there is limited flexibility to shift the training ammunition that DA Pamphlet 350-38 authorizes for record fire to other events they may consider important, such as having soldiers conduct shot grouping exercises.

Types of engagements. The second, and far more important, issue is that the live-fire qualification standards do not include all the types of engagements that are logically and doctrinally required in combat.

---

[6] In a grouping exercise, a soldier fires three to five rounds at a point on a target. The closer the strike of all the rounds are, the better the soldier's shooting skills. On a KD range, the soldier shoots at a specific point on a target at specified ranges, and feedback on exactly how far from the aim point the round strikes is provided to the soldier and coach. Both grouping exercises and KD shooting give the soldier and coach effective feedback to improve shooting techniques.

[7] A weapon is "zeroed" when the point of aim and point of bullet impact on the target are identical at a given range.

Rifle record-fire qualification requires the engagement of frontal, stationary targets arrayed in a 16 meter–wide lane and at ranges from 50–300 meters, with eight targets being beyond 200 meters.[8] Of the 40 targets, 30 are single target engagements with between three-second exposures (for the target at 50 meters) and eight-second exposures (for the target at 300 meters). Two targets appear simultaneously in 10 of the engagements, with between six-second exposures (for the targets at 50 and 100 meters) and 12-second exposures (for the targets at 150 and 200 meters). Twenty of the engagements are from the prone-supported or foxhole-supported positions, ten are from the prone-unsupported position, and ten are from the kneeling-unsupported position. The engagements from the kneeling position are between 50 and 150 meters.

- This review shows that many types of combat engagements are not included.
- When attacking, enemy soldiers would be moving—often in short rushes—but there are no engagements of this type.[9]
- Quick "reflexive" engagements are frequent and often occur at short range. While these are included in ARM, they are optional and not authorized in DA Pamphlet 350-38 for soldiers in other than infantry positions.[10]
- In the offense, engagements from a standing position or from standing to kneeling are frequent, and again there are no qualification engagements of this type.[11]
- When defending, tactically proficient enemy soldiers who are stationary will expose themselves as little and as briefly as possible. From this perspective, the stationary engagement times in the record fire are relatively long.[12]
- The fields of fire on the record-fire course are very narrow compared to many combat requirements.[13]
- The M16/M4 is designed to fire three-round bursts, but there is no exercise of this type of engagement.
- The ranges are level, whereas shooting at elevated targets or from elevated firing positions is common during urban operations and is frequent in Afghanistan.
- Soldiers normally use night vision goggles for night operations and these can be very effective in conjunction with the laser AN/PEQ-15 Advanced Target Pointer/Illuminator Aiming Light. However, unlike the case for machine guns and SAW, there is no rifle qualification requirement for engagements with night vision goggles.

---

8 See FM 25-8, *Training Ranges*, May 2010, for a full description of the standard record-fire rifle range.

9 See the task "Move Under Direct Fire" in Soldier Training Publication (STP) 21-1 SMCT, *Soldiers Manual for Common Tasks Warrior Skills Level 1*, June 2009. It describes the technique of using short 3–5-second rushes when moving under fire, and even this technique is used only selectively compared to high or low crawl techniques.

10 This type of engagement can be important for support-unit soldiers as well. For example, imagine a support unit in a rear area support facility being subject to an enemy raid at night.

11 See TC 7-9, *Infantry Live-Fire Training*, September 1993. During offensive operations, a rifleman will often have to immediately return fire from a standing position or go to a kneeling position because fields of fire from a hasty prone position are often limited.

12 See discussion of defensive positions in FM 3-27.8, *The Infantry Platoon and Squad*, March 2007.

13 For example, a rifle marksmanship instructor at the Maneuver COE described the "10 to 2" (on the face of a clock) as a rule of thumb for typical combat sectors of fire. The sector of fire on a standard automated record-fire range is a small fraction of this, 20 meters wide at 300 meters (see TC 25-8, *Training Ranges*, May 2010c). This same issue also exists on SAW and machine gun qualification ranges.

These points do not imply that the current record-fire standards are wrong; they exercise important rifle marksmanship skills. But they do not validate proficiency of many other critical combat-rifle engagement skills. The importance is that modified techniques and skills are needed for many of these engagements.

A comparison of machine gun and SAW qualification standards with combat requirement leads to similar findings.

Both machine gun and SAW qualification require the completion of three tables. Table I is a short-range (10-meter) exercise that is done on the EST. Table II is a day live-fire event, and Table III is a limited visibility live-fire event, normally done with night vision goggles.

Table I, EST qualification, requires the gunner and assistant gunner to exercise their ability to engage targets using traverse and search techniques by engaging first a series of linear targets in depth and then a set of area targets with width and depth.[14] A five to seven–round burst is fired at each target, and the shooter gets a point for each target hit (up to seven per target). A total of 63 of 91 rounds must hit the target for a qualifying score.

The Table II, day live-fire qualification, involves seven engagements of point targets for both the machine gun and SAW. Four of these engagements involve single targets, two have two targets, and one has three. The ranges are longer for the machine gun (400–800 meters) than for the SAW (100–400 meters). The SAW requires one engagement while wearing a protective mask, but this is not the case for the machine gun. Engagement times are 10–30 seconds for the single targets, 30–40 seconds for the two-target engagements, and 45 seconds for the three-target engagement. Seven of the 11 targets must be hit to qualify. The machine gun may be fired from either a tripod or bipod.

Table III, night live-fire qualification, is similar to Table II in that it involves seven engagements of point targets and has the same mix of single and multiple targets. However, the ranges are shorter.

The lack of a full range of combat engagements is even more of an issue for machine gun/SAW qualification than for rifle qualification.[15] Key considerations include the following:

- The standard machine gun/SAW ranges have sectors of fire with limited width (about 10 degrees).
- A competent enemy will not willingly stay stationary and exposed while being shot at, yet these are the types of engagements exercised.
- Machines guns and SAW have a major area-fire role, but area-fire skills are only exercised during EST training. Furthermore, this training is mechanical and does not present the gunner realistic target arrays.
- Machine guns are normally employed in pairs, but there are no structured exercises of these types of engagements.[16]

---

[14] *Traverse* is changing the point of aim to move the strike of the bullet horizontally. *Search* is changing the elevation of the strike. Engaging an area target requires traverse and search. A machine gun on a tripod has a traverse and elevation mechanism to aid in performing area fires.

[15] See FM 3-22.68, *Crew-Served Machine Guns 5.56-mm and 7.62-mm,* "Chapter 5: Combat Techniques of Fire," July 2006, for further discussion of combat standards.

[16] See "alternating fire" and Appendix A, "Machine Gun Employment," in FM 3-21.8, *The Infantry Rifle Platoon and Squad*, March 2007.

- In the offense, over-watching machine guns often "march" their fires ahead of assaulting fires. Overhead fire is another advocated technique, but these are not exercised or feasible on standard live-fire ranges.

Collective live-fire exercises could exercise some of these types of engagements, but TC 7-9, *Infantry Live-Fire Exercises*, does not include any doctrinal guidance on what types of small-arms engagements to include in these events. They are rightfully described as being mainly maneuver and command and control events using the Mission Training Plans (MTPs) Training and Evaluation Outlines (TEOs), which emphasize tactical and leader tasks in their evaluation. In any event, given the targetry and instrumentation on most standard, collective live-fire ranges, evaluation and feedback on individual and crew weapon engagement success would be very difficult to implement fully.[17]

Examination of Small-Arms Training Strategies in Other Nations' Armies and the U.S. Marine Corps

To put the Army's small-arms strategies in perspective, we examined the small-arms strategies of several other nations and the USMC. The examination focused on rifle training strategies but included those for machine guns as well.

To conduct this examination, we contacted the Maneuver COE's Australian, British, Canadian, French, and Republic of Korea liaison offices and the USMC's Weapons Training Battalion at Quantico, Virginia, and asked for information on their small-arms training programs. The information came back in different forms. In some cases, documents were sent; in other cases, we were sent emails describing their programs.[18] In all cases, we discussed these programs in detail with our contacts. While the data we collected were not sufficient for a structured, systematic comparison, we were able to identify common elements of these programs and compare them with the Army's small-arms training strategies to identify key differences. Key differences are summarized in Table 4.2.

This review shows that the programs of the other nations and USMC are similar to those of the Army in that they devote considerable resources to small-arms training. However, many specifics of these programs differ from the Army's. In many respects, they are more like each other than they are like the Army's programs. While it is not possible to compare the outcomes (e.g., better combat shooters), the results of this comparison provide a reasonable set of inputs for the Army to consider in its efforts to improve its small-arms training strategies.

Ranges and Targetry

In addition to training ammunition, a second key category of resources needed to support small-arms training is ranges and targetry. The Army uses the Army Range Requirements Model (ARRM) to support the process of assessing the adequacy of this type of support. This

---

[17] The Army is developing improved capabilities in this area, but fielding a capability to make this available on a scale to support current small-arms strategies is not possible in the near or mid-term.

[18] The main documents we received were Marine Corps Order 3574.2K, *Marine Corps Combat Marksmanship Programs*, August 2007; the British Army's Army Code Number 71810, *Operational Shooting Policy*, February 2011; and Australian Army's Defense Instruction (Army) Admin 89-1, *Army Small Arms Weapon Proficiency Policy*, December 2008. We did not get formal documents from the French and Korean liaison officers, as these had not been translated into English. But they did write summaries, answer questions and email queries, and discuss their nations' strategies in depth. The same was the case for the Canadian small-arms strategies.

Table 4.2
Key Differences Between Army Small-Arms Strategies and Those of Other Nations and the USMC

Difference	Remarks
Annual rather than semiannual qualification	True for all except Republic of Korea Requires qualification three times a year because of constant threat
Structured, required, resourced preparation for qualification	Includes procedural (e.g., assembly/disassembly) and marksmanship (e.g., grouping) events Tested requirement and prerequisite to qualification with allocated ammunition
Structured use of simulators	Simulator exercises a prerequisite Exception is USMC, but its strategies are simulator supported
CBRN engagements not a separate event	CBRN is an engagement in an event rather than a separate event
Secondary sights not a separate event	Selected engagements in events rather than a separate event
Short-range engagements and moving targets required	Included engagements in structured programs
Less collective live fire	Fewer events Lower echelons (most only to squad level) Exceptions are the French army, which has a two-week, company-level live-fire event every two years, and the USMC, which has an extensive LFX at 29 Palms before deployment
School Trained Small-Arms Trainers	Taught in either stand-alone courses or as a major part of required professional development courses
Machine gun MOS	All had an MOS and career-development path

model is an integrated planning tool. Based on force structure data from the Army Stationing and Installation Plan (ASIP) and DA Pamphlet 350-38's strategy requirements, ARRM calculates each installation's throughput capabilities and requirements.[19]

With the assistance of the Army Training Support Center staff overseeing these resources, we reviewed ARRM data for several AC troop installations. This review indicated that their small-arms range capabilities were adequate to support the current individual and crew small-arms qualification requirements. However, there were capacity limitations in terms of meeting some ARM requirements (examples include moving-target engagements and extended-range engagements for SDM training).

Also, collective live-fire range capabilities fall far short of being able to support required qualification (e.g., quarterly squad/platoon LFX) and recommended events (e.g., annual CALFEX). This shortage could be a major issue considering that many unit commanders would want their units to go through a practice dry-fire run before going through with a maneuver exercise with live ammunition.[20]

While ARRM provides good visibility of the quantity of ranges needed to support currently approved DA Pamphlet 350-38 requirements, it does not provide visibility of many quality issues—for example when ranges do not support important combat requirements currently

---

[19] See Department of the Army, *Army Range and Training Land Strategy*, February 2004.

[20] For example, TC 7-9, *Infantry Live-Fire Exercises*, describes the need for a crawl-walk-run approach to LFX preparation and suggests a leader Tactical Exercise Without Troops (TEWT), a walk through, and a Multiple Integrated Laser Engagement System (MILES) exercise before an LFX, but says this training does not need to be done on the actual LFX range.

approved in DA Pamphlet 350-38. A key example in this regard can be seen in SDM training. Based on combat experience in Afghanistan, the Army has doctrinally recognized the need for having the ability to engage targets beyond 300 meters and has issued weapons and sights to support this requirement. However, because there is no specific DA Pamphlet 350-38 requirement for this training, the ARRM does not show range shortfalls to support this type of training. Similarly, other important range training quality factors not directly visible in ARRM nor used to determine its adequacy ratings include moving targets, wide sectors of fire, area engagements for machine guns, heated targets to support engagements with thermal sights, and automated feedback for known distance ranges.[21]

The inability to support critical combat requirements particularly affects collective live-fire training, in which the combination of limited range space and large service-round down-range danger areas means that, in most collective LFX, the all-around security and maneuver that would be normal in actual combat cannot be effectively exercised.[22] For this reason, FORSCOM has been urging the Army to change its policies for procuring small-arms training ammunition to include a far higher percentage of Short Range Training Ammunition (SRTA).[23] Having more of this type of ammunition would enable the Army to have more realistic training.

Another collective live-fire issue is range instrumentation capabilities. As described previously, there are no Army-wide engagement standards for collective live-fire exercises. But current range automation capabilities could not support such standards if they were established or if desired by local commanders. For example, the assessment of effectiveness would benefit from knowing what shooter hit which target. Making these judgments using human trainers, while possible to some degree, would be difficult, would require a lot of manpower, and would be less accurate than instrumented scoring. While some plans exist to improve instrumentation in this area, almost no real capability is available at this time, and there is no prospect of funding to support the large number of collective live-fire exercises outlined in DA Pamphlet 350-38.

We make these points not to say that the Army should necessarily allocate the high level of resources needed to remediate these issues fully to ranges and targetry programs. But the Army should look at alternatives to live fire and improvements (such as procuring more SRTA and less service ammunition and using EST to practice moving-target engagements) to more completely exercise and test critical engagement skills. Moreover, shortfalls and their possible consequences should be more directly visible to Army decisionmakers.

EST Availability

As we have seen, the EST also plays a key role in small-arms training strategies. For example, it supports the semiannual qualification requirement for CBRN and unassisted night fire for the

---

[21] The Army has recognized the benefits of automated feedback and fielded a Location of Miss and Hit (LOMAH) for automated feedback on some Initial Entry Training ranges to reduce the manpower support costs needed to provide immediate shooter feedback.

[22] For example, DA Pamphlet 350-38 currently authorizes only 150 Short Range Training Ammunition (SRTA) rounds per year for Infantry ARM.

[23] Other reasons support the need for SRTA to allow the use of current range capabilities. The range of 5.56mm service ammunition has increased, and this exceeds the usable range fans of many small-arms ranges. As a result, some firing points cannot be used or require special safety procedures.

rifle and the 10-meter tables for the SAW and M240B machine gun. Additionally, the strategies for these weapon systems call for the use of EST to support quarterly PMI. The EST also supports the semiannual qualification requirement for the M9 pistol (CBRN and unassisted night fire) and supports its quarterly PMI. Thus, the EST is required for the individual weapon training of virtually every soldier in a tactical unit.

The EST supports PMI for almost all other small arms (and other weapons such as the AT4 light anti-tank weapon) as well. It also plays a major role in the Maneuver COE's emerging training strategies for mounted unstabilized weapon systems (e.g., .50-caliber machine guns mounted on trucks for self-protection). EST capabilities also support squad-level collective weapon engagement and "Shoot, Don't Shoot" training.

From many perspectives, the EST has great potential to provide additional benefit to small-arms training. In addition to its obvious potential to support PMI, it has the especially important potential to exercise and test small-arms skills that are not currently exercised or tested, including moving and evasive enemy targets and area fire for machine guns, in that the EST would make it possible to develop realistic scenarios to exercise and validate such skills.

As outlined in Chapter Three, the issue with regard to leveraging EST to enhance small-arms training is that current EST fielding is not sufficient to support current strategy requirements, let alone expand its use to support training additional requirements.

However, in spite of the fact that the current number of ESTs is not sufficient to support current small-arms strategies, current usage indicates that it may be possible to expand the role of the EST. Several TRADOC EST staff members stated that the recent usage rates have been low—reportedly in the neighborhood of 30 percent of the available hours.[24] While this low usage rate is almost certainly a result of the high rate of operational deployments, to some degree it must also indicate that the units are not following weapon training strategies, including qualification requirements.

Findings

The following key findings arise from this examination of small-arms strategies and the resources necessary to support them:

Many Critical Individual/Crew Small-Arms Tasks and Skills Are Not Included in Qualification Standards or Fully Supported in the Strategies

Many small arms skills needed by all soldiers who deploy to a combat theater are not qualification requirements or even included as skills to be trained in current DA Pamphlet 350-38 strategies. Only infantry soldiers are authorized additional training ammunition for some of these skills in ARM, but ARM is not a qualification requirement.

[24] We reviewed EST usage at Fort Carson, Colorado. Fort Carson, which has six 10-station EST sets on the installation, has about 24,000 assigned soldiers. Between February 2011 and January 2012, about 32,000 soldiers received EST training—a very low use of the capacity, even when taking into account that many Fort Carson soldiers were deployed or recovering from deployment during this period.

Most Individual/Crew Strategy Training Ammunition Is Allocated to Repetitive Qualification, but There Is Limited Support for Preparation Events/Activities

Except for infantry ARM, almost all ammunition to support individual/crew small-arms training is allocated to semiannual qualification or direct preparation for qualification. However, the resources to support the development of marksmanship skills to prepare soldiers for qualification are limited. No ammunition is allocated to grouping or similar engagement-skill development exercises. There are no institutional "train the trainer" programs to support effective preliminary marksmanship training, and simulator resources are insufficient to support quarterly PMI. This issue is especially acute for machine gun training given the lack of a Military Occupational Specialty (MOS) or Additional Skill Identifier (ASI) for these weapon systems and the fact that only familiarization training—at best—occurs in Initial Military Training.

Collective Live-Fire Exercises Account for a Large Percentage of Training Ammunition and Other Resources but Have No Engagement Standards and a Limited Support Structure

Collective live-fire exercises are the most frequent live events for infantry soldiers and account for most of their annual rifle-training ammunition authorizations and a large percentage of their automatic rifle and machine gun authorizations. Even for support units, the annual collective live-fire exercises account for a high percentage of the small-arms training ammunition authorizations. Additionally, live-fire exercises require extensive unit and leader time to plan, prepare, and support execution.

However, there are no true engagement standards and only a limited range capability to directly support engagement skills assessment and improvement during these exercises. Also, the doctrinal support for planning and conducting these very complex exercises is limited and there is no institutional leader training in this area. Finally, the number of ranges on installations is not sufficient to support the number of exercises outlined in the strategies.

Conclusions

These findings lead to the following three conclusions:

Standards Should Change to Include a Fuller Range of Needed Combat Engagement Skills

This issue may be partially explained by the fact that small-arms strategies focus almost exclusively on what can be done on live-fire ranges and do not recognize the potential of the EST's virtual capability to train and validate engagement tasks and skills that are not practically feasible on current live-fire ranges. In this regard, an important aspect is developing engagement standards for collective live-fire training events so that the ammunition and time allocated to these events add to the development of small-arms engagement skills and the ability to apply them in combat situations.

Strategies Should Change to Better Resource Preparation Activities

The strategies of most other nations and the USMC have ammunition allocated to training events that develop basic marksmanship skills (e.g., shot grouping and KD exercises with downrange feedback) in preparation for qualification. The lack of a system to provide trained small-arms trainers to units is another major obstacle to effective preparation training.

Strategies Should Change to Align Better with Progressive ARFORGEN Readiness Processes

The DA Pamphlet 350-38 strategies do not support efficient, progressive ARFORGEN readiness. The quarterly requirement for an infantry squad/platoon LFX and a semiannual platoon/company LFX is especially questionable early in the ARFORGEN cycle given the considerations that the progressive readiness concept and that manning Aim Points do not require the Army's manning system to provide for the personnel stability needed for effective preparation in the Reset portion of the cycle.[25]

---

[25] See AR 525-29, *Army Force Generation*, March 2011. This is not to say that there could not be some collective LFXs earlier in the ARFORGEN cycle, but that they should not be a qualification requirement or gate.

CHAPTER FIVE

Improving Small-Arms Training Strategies

In this chapter, we outline considerations, principles, and specific directions the Army should consider in its ongoing efforts to make small-arms strategies more effective and efficient. This effort necessarily involved judgment on the part of the research team, but we sought and considered the input of a range of knowledgeable members of the small-arms training community, including trainers from the Army Marksmanship Unit, Special Forces small-arms trainers at Fort Bragg, and staff at the Maneuver COE who are currently in the process of developing revised small-arms training strategies. In general, what we outline aligns with the Maneuver COE's revision efforts and reinforces or supplements directions it is taking.

Considerations

A major consideration is that the Army is entering a period of major adjustment. It is transitioning from an era in which a large percentage of units were deployed and their training programs were almost exclusively focused on counterinsurgency and stability operations to an era requiring a broader range of operational requirements, in which specifics of the training environment are still emerging. Some key considerations that affect directions that small-arms training strategies should take include the following:

Operational requirements reinforce the continuing need for small-arms engagement skills. There is no well-defined central threat. However, a reasonable assumption is that a competent future enemy will not try to fight U.S. ground forces "head on" and conventionally. The high relative conventional capabilities of the Army were clearly evident during operations in Afghanistan and Iraq, and continued technological advances (command and intelligence information systems, unmanned aerial aircraft, armed drones, etc.) have increased these advantages. Instead, a future "hybrid" threat will more likely try to fight using guerrilla-type tactics. Such tactics include avoiding engagement unless the circumstances are favorable and dispersing and hiding when they are not, fighting in urban and other complex terrain, striking support facilities and units rather than combat units, using ambush tactics, and similar tactics and techniques to survive and prolong the conflict. This means that small unit actions and the small-arms skills of all soldiers, including precision fire for successfully engaging the enemy while avoiding friendly force and noncombatant casualties, will be of major importance.

ARFORGEN processes are changing. While requirements for combat unit deployments are expected to decrease, the need to support a range of ongoing CCMD/ASCC requirements and possible contingencies with a smaller AC force has caused the Army to begin revising its

AC ARFORGEN process from a 36-month cycle to one of 24–27 months.[1] Effective small-arms strategies must be shaped to develop critical small-arms skills during the shortened Reset and Train-Ready phases and to sustain them during the Available phase.

Maneuver Combat Training Center (CTC) rotation design and timing are evolving. The Army is developing "Decisive Action" maneuver CTC rotational training to prepare BCTs for the range of possible threats. The redesign includes live-fire exercises that will remain important rotational activities, but their specifics in terms of small-arms tasks and skills are still being developed. The timing of maneuver CTC rotations in the new, shorter ARFORGEN cycles is also being examined. The specifics of this timing and the scope of live-fire events will affect units' home station live-fire training strategies. These must be integrated for the best possible mission readiness. Home station weapon training should first prepare the unit for the CTC live-fire events and, subsequent to the rotation, train additional mission-required small-arms tasks and skills.

Manning strategies have not been defined. Key elements of weapon training strategies are the timing and frequency of weapon training activities. Unit personnel fill rates and the level of personnel turnover heavily influence these elements. Some weapon training should be deferred until the leaders, individuals, and crews that will man the unit and weapon systems at deployment are reasonably stable.

Designated weapons are a good example of where manning strategies have an impact on weapon training strategies. M240B machine guns in BFV infantry companies are designated weapons; that is, they do not have a crew assigned to man the machine gun as a primary mission. While there are some benefits to continually maintaining trained and qualified designated crews, from the perspective of making the best use of unit training time, ammunition, and range resources, it makes sense to train the crews "just in time," or shortly before they enter their deployment windows, rather than requiring continuous semiannual qualification.

Personnel stability also affects collective live-fire training. Army doctrine recommends a crawl-walk-run approach to preparing for collective live-fire exercises, and the unit time needed to plan, prepare, and execute these events is substantial, as are the range and training ammunition resources required to support them. Performing these exercises before leadership and key weapon position manning are stable is likely to provide limited deployment readiness relative to the costs and efforts involved.

Conversely, the earlier a reasonable level of unit stability can be achieved, the earlier progressive weapon skills training can begin and the higher the skill level that can be achieved with equivalent amounts of training. As a result, the requirement to retrain crews will be lower, and less effort will be needed to sustain collective live-fire proficiency.

The issue for weapon training strategy developers is that the timing and levels of personnel fill rates and the levels of personnel turnover across the ARFORGEN cycle have not been determined.[2] While the shorter ARFORGEN cycle could allow greater stability, the degree to

---

[1] General Ray Odierno stated the need for the Army to be capable of "anywhere and anytime" support of CCMDs, to have regionally aligned forces, and to adjust ARFORGEN to meet these needs at the February 2012 Association of the Army Winter Symposium. These requirements are also outlined in the *2012 Army Posture Statement*, February 2012a.

[2] See AR 525-29, *Army Force Generation*, which outlines a concept called ARFORGEN Focused Manning. "The arrival of Soldiers assigned to brigade size units will be synchronized based on unit's deployment latest arrival date (LAD), major training exercises, and unit's last redeployment date. Replacement personnel will be assigned utilizing the Army's Individual Replacement System. These replacements will likewise be assigned and synchronized to the unit's operational cycle. The Deputy Chief of Staff, G-1 (DCS, G-1) establishes manning policies in support of ARFORGEN aim points."

which stabilization policies will apply below battalion command level is not yet clear.[3] Current policies aim at high personnel readiness at the time of deployment but do not establish specific manning targets for Aim Points along the cycle, and they state that the individual replacement system will be used to maintain personnel fill.

Unit time to train will remain the key constraint. While lessened requirements for operational deployments will have a positive effect on unit training time, this is somewhat offset by the requirement to train on a wider range of tasks, skills, and conditions. Time was the biggest constraint on unit training before deployments to Iraq and Afghanistan began, and this constraint will remain.

Training program funding will be difficult. As pointed out throughout this report, the likelihood of getting additional training resources to respond to increased requirements is low, and even avoiding reductions will be a challenge to the training community.

Weapon training strategies will evolve, but the specifics are not clear. The overall conclusion is that, while reasonable directions for revising small-arms and other training strategies can be defined in the near-term, the specifics will evolve, and changes, likely major ones, will be required as the elements outlined above solidify and their effects become clearer.

Principles

Based on the small-arms training findings, conclusions, and considerations, we developed some broad principles to guide the development of specific directions for improving small-arms training strategies. These are outlined in Table 5.1.

Train and Qualify All Soldiers on the Critical Individual and Collective Tasks and Skills Necessary for Self-Protection

Future enemy forces are likely to use tactics that include attacking support units throughout operational theaters.

Require Advanced Qualification Levels and Additional Skills for Soldiers Who Need Small-Arms Skills for Mission Accomplishment

Soldiers in units with the mission of finding and defeating enemy forces should be qualified on all tasks and skills critical to accomplishing that mission.

Leverage the Full Range of Modalities to Train and Qualify as Efficiently as Possible

Current qualification requirements focus on what can be done on live-fire ranges. In current small-arms training strategies, the EST is used for basic tasks or preliminary training, but its capabilities are far greater.

---

3 The Chief of Staff of the Army has stated that changes of command would be done at the end of the Available period to facilitate stability through the end of the next Available period, but further stabilization policy goals have not been yet been established.

Table 5.1
Recommended Principles for Improving Small-Arms Training

Principle	Justification
Train and qualify all soldiers on critical self-protection individual and collective tasks and skills	All soldiers need self-protection skills in current operational environment
Require advanced qualification levels and additional skills for soldiers who need small-arms skills for mission accomplishment	Soldiers with missions of finding and defeating the enemy require additional skills beyond self-protection
Leverage the full range of modalities to train and qualify as efficiently as possible	The EST and other simulators have greater capabilities than outlined in current small-arms training strategies
Make training and qualification progressive and align qualification timing to Available pool readiness requirements	A progressive strategy could better align to the Army's ARFORGEN doctrine
Establish and provide resources for structured training/skill to prepare for qualifications	Current strategies do not provide the resources or requirements to test and develop the underlying skills required for qualification
Formally develop unit leaders with effective small-arms trainer skills	Implementing enhanced training strategies will require expert trainers
Assess, evaluate, and continuously improve small-arms strategies	A continuously evaluated training program would be better to adapt to new operational requirements for small-arms skills

Make Training and Qualification Progressive and Align Qualification Timing to ARFORGEN Readiness Requirements

The current strategy is repetitive rather than progressive. It requires semiannual validation of a subset of basic marksmanship skills, while not providing for the development of a full set of small-arms combat-critical tasks and skills.

Establish and Provide Resources for Structured Training/Skills to Prepare for Qualification

The current strategies test basic marksmanship without requiring or providing the resources for the development and testing of the basic marksmanship skills required for qualification, such as rifle live-fire shot grouping exercises, or a required PMI gate before qualification.

Formally Develop Unit Leaders with Effective Small-Arms Trainer Skills

Implementing progressive training strategies that include coached dry-fire drills, grouping exercises, effective machine gun area-fire engagements, fully leveraging the EST for preliminary and advanced marksmanship training, and conducting effective collective live-fire exercises would benefit greatly from systemically ensuring unit commanders have formally trained small-arms trainers. The U.S. Army is alone in not having institutional small-arms "train the trainer" courses for unit leaders.

Assess, Evaluate, and Continuously Improve Small Arms Strategies

Because the small-arms and other training strategies defined in the near term will need refinement and likely major changes over the longer term, a key need is having a structured evaluation and assessment program to support continuous revision and improvement.

Specific Directions for Small-Arms Training Strategy Improvement

In this section, we outline the specific directions we developed for applying these principles. These directions are summarized in Table 5.2, which displays the principles in the left-hand column and explanations of each direction in the right-hand column.

Develop Enhanced, Prescriptive Individual and Crew Standards Using the Full Range of Virtual and Live-Fire Modalities

The first and most important effort should be to develop small-arms engagement tasks and standards that reflect critical combat requirements. The basis for additional individual qualification tasks for rifles (moving targets, reflex fire, etc.) already exists in the ARM tables, and similar tasks could be developed for machine guns and the SAW. Moreover, the Maneuver COE is continuing to examine improvements to rifle training strategies to better prepare soldiers for combat.[4] An important resource to support these efforts is the large amount of small-arms combat experience currently available among its junior leaders.

As discussed above, the need for increased standards applies to all soldiers. For example, only infantry soldiers currently have reflex engagements authorized in the DA Pamphlet 350-38 strategies. This standard ignores the fact that all soldiers must be able to protect them-

Table 5.2
Directions for Improvement

Principles	Directions for Improvement
Train and qualify all soldiers on critical individual and collective tasks and skills necessary for self-protection	Develop enhanced, prescriptive individual and crew standards using the full range of virtual and live-fire modalities
Require advanced qualification levels and additional skills for soldiers who need small-arms skills for mission accomplishment	Formalize live-fire requirements for deployment
Leverage the full range of modalities to train and qualify as efficiently as possible	Improve the small-arms engagement skill benefits of collective live-fire exercises
Make training and qualification progressive and align qualification timing to Available pool readiness requirements	Develop "resource neutral" progressive, descriptive strategies to achieve combat standards
Establish and provide resources for structured training to prepare for qualification	Identify, develop, and field cost effective training capabilities to support basic marksmanship skills
Formally develop unit leaders with effective small-arms trainer skills	Develop and institute formal "Train the Trainer" programs
Assess, evaluate, and continuously improve small-arms training strategies	Implement a structured process to support small-arms training strategy development and evolutionary improvement

[4] See Jean L. Dyer, Peter S. Schaefer, Martin L. Bink, David R. James, Richard L. Wampler, and Michael D. Dlubac, *Soldier Performance on a New Marksmanship Course of Fire*, Alexandria, Va.: U.S. Army Research Institute, Research Report 1924, June 2010. This report outlines the Army's development of a new rifle training course of fire that modified the current record-fire table to more closely represent combat conditions in Iraq and Afghanistan. Additions to this new course of fire included reacting to simulated malfunctions and shooting from a kneeling position behind a barricade. This report also includes an appendix that presents rifle marksmanship qualifications going back to 1940 and discusses how these have evolved.

selves. Soldiers in a support facility, for example, must be able protect themselves and their facility in the event of an enemy raid.

Combat arms soldiers would logically have additional tasks and higher standards. For example, infantry soldiers should have longer-range engagements in their rifle qualification tables. While the machine gun qualifications for support units could center on defensive engagements, those for infantry should also include offensive tasks and skills, such as providing over-watching fires in support of maneuvering elements.

A key consideration in developing these tasks and standards is ensuring they are critical and reasonable—that is, that they are truly key to combat success. The bar can be set high but should not be so high that the typical soldier cannot achieve success with a reasonable preparation program. This means that, before prescribing Army-wide standards, the standards developed by proponent subject-matter experts should be tested and validated in actual units executing normal training programs.

An important shift in current qualification approaches would be to use the EST to train and qualify critical engagement tasks that are not feasible on live-fire ranges. There is only so much ammunition the Army can allocate to small-arms training. Further, training on live-fire ranges requires far more time for units to plan, prepare, and coordinate than do EST exercises. Moreover, the available targetry and instrumentation capabilities of ranges for many of these types of engagements are constrained and would require extensive resource allocations to expand. Key examples are moving targets, wide sectors of fire, automated downrange feedback, and automated area-fire engagement evaluation. Likewise, an EST scenario can, in many respects, provide engagements that are more realistic portrayals of combat targets than are provided by standard live-fire targets. This is not to argue that engagements in the EST are better. The EST's indoor engagements with subdued noise and recoil and stationary positions also lack important elements of realism. But the EST and live-fire ranges can be complementary in terms of exercising a fuller range of combat tasks and conditions.

Improve the Benefits of Collective Live-Fire Exercises for Small-Arms Engagement Skills

An important need is improving collective live-fire standards. Currently, there are effectively no small-arms engagement tasks or standards for collective LFX, even though these exercises represent a major portion of the annual training ammunition allocations. Instead, TC 7-9, *Infantry Live-Fire,* refers to the TEOs in infantry unit MTPs.[5] However, these TEOs focus on maneuver and leader command and control tasks and standards. They do not include guidelines that would identify specific live-fire engagement tasks for inclusion or promote effective weapon skills evaluation.

While commanders need flexibility for these types of events (e.g., the capabilities of collective live-fire ranges vary considerably across installations, and units have different training requirements and training programs to meet them), a greater degree of structure and guidance could provide for greater engagement skill benefit. For example, there could be guidance on target exposure times and presentations. Engagement-specific checklists to supplement tactically focused TEOs could be developed (e.g., what percentage of targets were engaged/hit; how

[5] TC 7-9, which provides guidance and examples to help commanders plan, execute, and evaluate LFX, was published in 1994. Since then, the Army has stopped producing MTPs, but the Maneuver COE published TC 3-21, *Infantry Rifle Company Collective Task Publication*, September 2012b. Like predecessor MTPs, it contains TEOs that outline conditions, standards, and subtasks for each critical task, but specific engagement tasks and standards are not included.

many machine gun, automatic rifle, and rifle rounds were fired compared to target hits; was area fire effective; did each weapon crew and soldier in the organization identify and engage targets in their sector).

Planning, setting up, and executing realistic live-fire exercises are both difficult and complex. This is an area in which the Army has learned much since TC 7-9 was published in 1993. Collecting and documenting this learning could support greater Army-wide benefit for future exercises.

A reasonable approach to supporting collective live-fire improvement would be to involve the maneuver CTCs and the Maneuver COE in a joint effort to develop and institutionalize improved techniques, procedures, and guidelines.

Formalize Live-Fire Requirements for Deployment

Once the combat-related individual, crew, and collective small-arms training standards have been developed and validated, they should become formal components of the Army's process for declaring a unit ready for deployment and prepared to enter the Available pool. For example, a standard might be that 90 percent of the unit's soldiers have to be fully qualified on their assigned weapon, and that 85 percent of all machine gun crews must meet qualification requirements.

Develop "Resource Neutral," Progressive, Descriptive Strategies to Achieve Combat Standards

A major conclusion detailed in Chapter Four that arose as a result of our examination of current Army small-arms training strategies, those of other armies and the USMC, and discussions with small-arms experts was that Army small-arms training strategies should include structured preparation to develop the fundamental shooting skills to support success on combat-focused qualification events. The current small-arms qualification tables test a portion of needed combat skills but lack the downrange feedback needed to support basic marksmanship skill improvement. Also, as we saw in Chapter Two, most of the Army's other weapon training strategies have preparatory tables with allocated training ammunition and structured gunnery skills tests that are gates to qualification. For example, tank and BFV crews are required to conduct COFT exercises that progressively train specific gunnery skills, and crews have to reach prescribed levels as a prerequisite to live-fire training. Applying a similar concept to small-arms training would make sense, and, indeed, the Maneuver COE is developing such an approach now.

In the current fiscal environment, the option of adding events and the ammunition to support them to current small-arms training strategies is not a reasonable expectation, and if preparation live-fire gates are added, the resources they require would need to be offset. Indeed, the current guidance to the Maneuver COE states that revised weapon training strategies must be "ammunition neutral."

Two directions could be taken to allow small-arms training strategies to achieve more complete qualification standards and provide for resourced preparation events within current training ammunition authorizations. The first would be to reduce the frequency of qualification events to align them with a progressive, ARFORGEN readiness approach and placing some, especially collective LFX, toward the end of the Train-Ready phase. The second, which we discuss in greater detail later, would be to emphasize the use of simulators, dry-fire exercises, and effective coaching as components of preparation training.

Figure 5.1 shows one way the concept of reduced ARFORGEN qualification requirements and progressive small-arms readiness development could be reasonably applied to the ARFORGEN cycle of a non-deploying CEF AC BCT. The application illustrated is feasible, but other options are possible.

This approach would have overlapping "Build," "Qualify," and "Sustain" phases. The Build phase has activities that progressively build readiness to the level at which the unit is fully ready for deployment. It starts at the beginning of the Reset period and continues to the beginning of the Available period. Small-arms Build training during the Reset and Train-Ready periods could include individual/crew live, simulator, and dry-fire exercises and collective drill and force-on-force training.

Units in the Train-Ready pool have two major Qualify gates. The first (GATE 1) would be to meet the requirements to participate in the maneuver CTC live-fire exercises.[6] For small arms, this means individual and crew qualification as prescribed in DA Pamphlet 350-38. This could also mean a squad and platoon LFX, with an option for a company-level CALFEX.

The second major Qualify gate (GATE 2) would be to meet the full set of deployment requirements for individual weapon, machine gun, and SAW qualification before entry into the Available pool. Newly assigned soldiers and crews would need to qualify and, depending

Figure 5.1
One Way to Apply a Progressive Small-Arms Readiness Approach to a CEF AC BCT Readiness Cycle

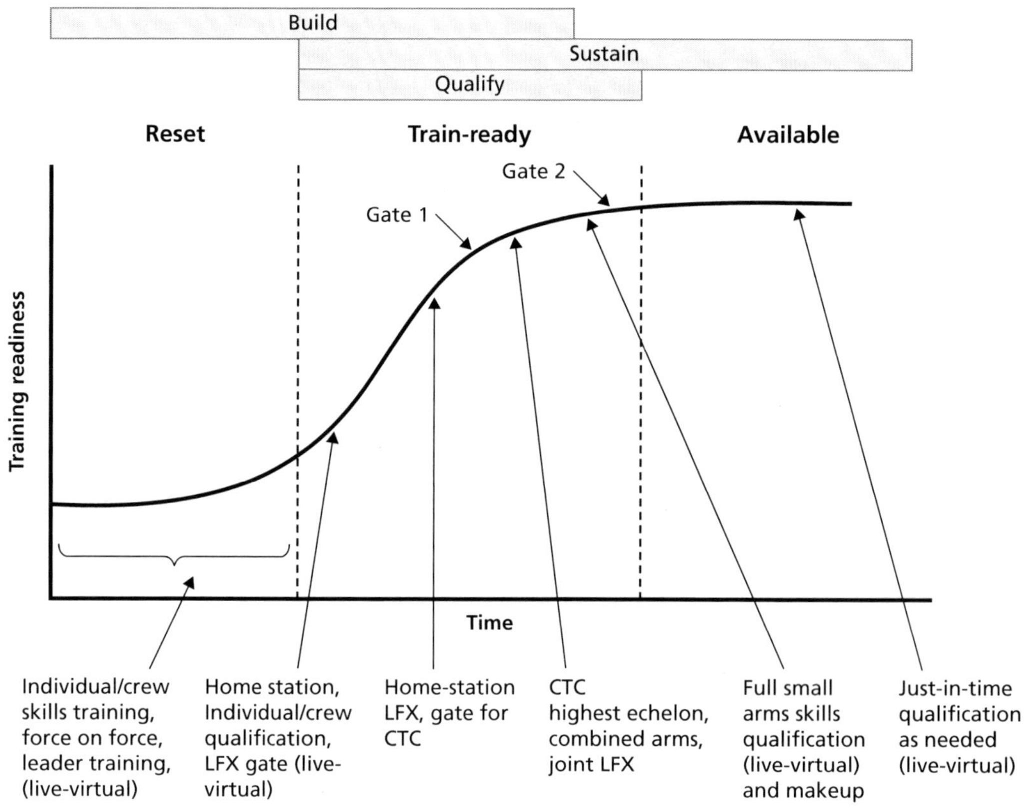

RAND *RR448-5.1*

[6] See FORSCOM Regulation 350-50-1, *Training at the National Training Center*, January 2010.

on the timing and type of training in the maneuver CTC live-fire exercises and the turnover in squads and platoons after the CTC training, additional collective live-fire exercises might be required to meet collective live-fire qualification requirements. For support units with a large number of designated crews and elements that did not conduct maneuver CTC live-fire exercises, a large portion of small-arms qualification might take place for this second gate.

Sustain-phase small-arms training begins after the first qualification and continues throughout the unit's cycle. It would mainly involve simulator, drill, and force-on-force exercises. However, sustainment training while in the Available pool would include a second full individual and crew qualification and likely a second set of squad and platoon LFXs.

The approach outlined shows how a progressive, ammunition-neutral training strategy could be developed that would support ARFORGEN readiness with fewer individual, crew, and collective live-fire qualification events. The reduced number of qualification events could allow for the redistribution of training ammunition from qualification to events such as grouping exercises. It could also allow greater training ammunition allocations to individual collective events. For example, each LFX is currently authorized 275 7.62mm machine gun rounds. The machine gun trainers we interviewed strongly believed this amount was not sufficient for an effective collective exercise.

Identify, Develop, and Field Cost-Effective Training Capabilities to Support Basic Marksmanship Skills

The EST has the capability to support training basic marksmanship skills. However, its fielding is limited and, as described in Chapter Three, cannot support much beyond the qualification requirements currently listed in DA Pamphlet 350-38. Moreover, a more important capability is arguably its use to train and validate skills on critical combat engagement skills not possible for most units on live-fire ranges. A positive consideration in this regard is that units do not seem to be using available EST capabilities, even for prescribed qualification requirements (e.g., the Unassisted Night Fire and CBRN Tables for the rifle), and this means that it might be possible to add EST advanced engagement qualification requirements (e.g., moving targets) without a practical impact on current unit training programs.

However, the combination of adding advanced qualification requirements and establishing EST training gates may result in increased usage, and current fielding may not support such enhanced strategies. At a cost of around one-half million dollars for a 10-station set, in the current budget environment, any major expansion of the number of fielded sets is not a likely option.

But other options could enhance the Army's capabilities to support enhanced basic marksmanship skills training to a degree that EST use could be shifted to focus more on advanced engagement skills. Most of the subject-matter experts we interviewed stated that dry-fire drills could be very effective. Also, as we have seen, the MACS and the LMTS have proven effective for training basic marksmanship skills, having the same benefits as dry-fire exercises but with the added advantages of more precise feedback; the potential for automated diagnostic tools; and, given relatively low costs, the opportunity for more "trigger squeezes" and more frequent training. Moreover, advancing technologies could expand these low-cost types of capabilities, not only for basic rifle marksmanship skills, but for automatic weapons and complex engagements as well.

Another fairly low-cost capability that could be considered would be adding Location of Miss and Hit (LOMAH) targetry on selected rifle ranges. At a cost of just over $100,000 for a

16-lane range, this would give a downrange feedback capability to zero and develop grouping skills at distances greater than on current 25-meter ranges.[7]

Develop and Institute Formal "Train the Trainer" Programs

Implementing these suggested directions would require a high level of training skills from the noncommissioned officer (NCO) supervisor to the staff level. Teaching and coaching engagement tasks and skills, the use of simulators, and the setting up of effective collective live-fire exercises all require a high level of small-arms expertise. As we have pointed out previously, the Army is alone in not having formal courses to train NCO small-arms unit trainers. This gap is even wider for machine gun training, for which other armies and the USMC think effective engagement and employment is important and specialized enough to make machine gunner a separate MOS.

Such training could be institutionalized in many ways. It could be added to infantry Non-Commissioned Officers Education System (NCOES) courses, as is done for Field Artillery NCOs. Another option would be to develop a standard course and support materials and allow individual installations or commands to establish local courses. We were told by training staff members at the Maneuver COE that many such local courses have been implemented. However, the option most supported by the Maneuver COE trainers we talked to would be to establish a unit small-arms master gunner course at the Maneuver COE, as is done for the tank, BFV, and Avenger. Included would be the award of an ASI and the establishment of "communities of practice" for keeping graduates updated and contributing to the exchange of best practices. A reasonable goal would be to have battalion- and company-level small-arms master gunners in military police, combat engineer, maneuver combined arms, and infantry units and battalion-level small-arms master gunners in other types of units.

The concept of a small-arms master gunner course has long been advocated by members of the small-arms community, but the resources necessary have never been provided and the concept has never progressed to implementation.

Implement a Structured Process to Support Small-Arms Training Strategy Development and Evolutionary Improvement

Developing and implementing more effective and efficient small-arms training strategies should involve near-term improvements, as well as long-term assessment and change.

Near term. The step of defining critical combat requirements (e.g., SDM, reflex/short-range engagements for the rifle, and area fire for machine guns) and adding them as qualification requirements is important and could be done in the near-term. The combat small-arms expertise in the Army is extensive and could readily support this effort. Also, implementation using the EST as a qualification modality for many of these tasks appears quite feasible.

Using military judgment to identify DA Pamphlet 350-38 events that can reasonably be reduced to make the addition of new qualification requirements resource neutral in terms of EST and small-arms training ammunition seems also possible without risk to readiness. For example, the need for semiannual CBRN and unassisted night fire on the EST could reasonably be reduced to annual requirements, especially in view of the low EST utilization rates.

In terms of training ammunition, the need to have the same number of engagements and frequency for night-sight practice and qualification as for primary sight and the high frequency

---

[7] The LOHAM cost estimate was provided by Army Training Support Center staff.

of required collective live-fire qualification exercises are areas where the ammunition supporting current strategies could be shifted to support additional combat requirements. An example would be adding a predeployment reflex-fire/short-range engagements qualification standard.

While identifying the needed tasks and skills and "bill-payers" could be done using military judgment, experimental studies would be required to validate and define the standards. The standards should set required skill levels that most soldiers can achieve with a reasonable preparation effort. An example of an area for experimental study would be to confirm that EST performance correlates to live-fire engagement performance on such tasks as moving-target engagement and machine gun area fire. Another hypothesis that should be tested before adaptation is that engagement with primary sights is sufficiently related to night and backup sight engagement that extensive practice and qualification events are not needed.

Military judgment can also be used to develop descriptive small-arms preparation strategies in the near term. But the establishment of prescriptive gates should be supported by experimental studies to verify that the gates and their standards truly predict qualification performance and that requiring them before qualification is either time efficient and reduces overall resource requirements or increases performance.

Establishing a "small-arms master trainer" course is another improvement that could be implemented in the near term.

Long term. While the Army can and should develop and implement better small-arms qualification standards and descriptive resource-neutral strategies for reaching these standards and should better align them with ARFORGEN readiness needs in the near term, a longer-term effort should be used to refine and adjust these strategies. As outlined in earlier sections of this chapter, many key operational requirements and training-environment factors are still emerging. As these become better defined, unit commanders will shape their training programs to meet requirements within the constraints of available resources, just as they did to make the major adaptations that were needed to train their units for operations in Iraq and Afghanistan. The Army should set up a program to monitor and understand how actual units' training programs evolve, how well they reach prescribed small-arms standards, how well they align with small-arms strategies (or why they differ), and in what ways they should be improved and better supported. This approach should be accompanied by structured observational studies to support analysis and improvement.

CHAPTER SIX

Developing and Assessing Options for Making Weapon Training Strategies More Efficient

In this chapter, we outline and examine several possible alternatives to the current M4/M16 individual record-fire qualification. Our purpose for developing and comparing these options is to show approaches the Army could use to develop and assess the viability of alternative courses of action, including different mixes of live and virtual means, to improve weapon training strategies.

We analyzed M4/M16 individual rifle record-fire training and qualification (and alternatives to these two linked events), but much of the underlying logic and results could apply to other rifle training events and other weapon system training strategies.

We selected the M4/M16 as an example case for several reasons. First, it expands on our detailed examination of small-arms training. Second, soldiers across most MOSs qualify on the M4/M16, and thus reducing training ammunition expenditures for this training could yield large savings. Third, a gunnery simulator, the EST, is available. Our analysis specifically includes simulator usage alternatives, so we deemed it best to focus on a weapon with a simulator that has already been developed and fielded.

Our analytical approach was qualitative in nature. It included reviewing prior rifle marksmanship studies and current rifle training strategies, gathering perspectives from trainers during interviews, reviewing EST capabilities as defined by capability documents, observing EST training at an installation, and interviewing those familiar with the simulator.

We identified three general options for reducing the amount of training ammunition used in future rifle training strategies[1]:

1. Reduce the number of rounds/engagements per event.
2. Reduce the number of events.
3. Reduce the frequency of the events.

These three options share a potential risk—reduced soldier capability—if not thoughtfully implemented. Because of this risk, we present approaches the Army could take to study these options before implementation.

While the three general options could apply to any weapon system, we developed options only for current individual record-fire rifle practice and qualification. Individual rifle record-fire marksmanship training involves several steps: (1) pre-marksmanship instruction (which includes use of the EST), (2) zero fire (18 rounds), (3) practice record fire (40 rounds), and (4)

[1] Combinations of these could be accomplished as well.

record fire (40 rounds).[2] For AC soldiers using weapons without advanced optics, this sums to 98 rounds every six months, or 196 rounds annually. The strategy stipulates that these steps are repeated for each weapon sight assigned to a soldier, but this examination is limited to strategies for soldiers assigned rifles equipped only with iron sights.[3]

Developing and Studying Options

We begin by expanding the three general options above to five by creating two versions of Option 2 and two versions of Option 3. After presenting an overview of the resulting five options, we provide a more thorough discussion of each. We also discuss research designs that the Army could execute to support informed decisions regarding the options presented.

Table 6.1 outlines a comparison of the baseline of 196 rounds per individual per year with each of the four alternative options.[4]

Table 6.1 summarizes the baseline and five alternate options (1, 2a, 2b, 3a, and 3b) in the columns and compares them with respect to the attributes contained in the rows: (1) means of reduction and number of rounds reduced in one year, (2) the specific changes to the current strategy, (3) our basis for recommending the option and its changes, and (4) possible approaches for further study of each option. Our purpose in presenting these options is to provide ideas and directions for approaches the Army could apply to develop feasible changes to weapon training strategies, not a prescription of whether or how to change these strategies.

Option 1—Shoot Fewer Engagements per Event

Comparing Option 1 with the baseline shows that it would provide an 80-round reduction in the number of rounds fired annually. This reduction results from firing fewer engagements. In this example, we present a case where one-half of the rounds currently in the baseline would be fired during each record-fire practice and qualification event.[5] Of course, shooting fewer engagements would only be warranted if doing so would not significantly reduce soldier marksmanship skills.

Some prior studies suggest that such a reduction in training ammunition without a decline in shooter proficiency may be possible. Notably, Hagman conducted an experimental study of 180 infantry soldiers going through rifle record-fire qualification training. He found a strong positive correlation between soldiers' scores shooting 20 rounds from a foxhole position

---

[2] Note that, while there are other rifle qualification requirements (e.g., collective LFX and unassisted night fire) and training events (e.g., ARM), rifle training strategies focus on record-fire qualification.

[3] The Army issues a variety of advanced rifle optics and is developing more. Some offer magnified views and others assist limited-visibility engagements.

[4] We assume the simplest case of 196 rounds as a baseline. This strategy would only apply to an AC soldier assigned a rifle with only an iron sight. For soldiers assigned additional sights, the baseline and other round counts could increase by a multiple of the number of sights assigned.

[5] The number of rounds required for zero in this option remains 18. Zeroing is not a training event, rather it is a procedure to ensure rifle sight is aligned to the strike of the bullet at a specified range by firing rounds and adjusting the sights to the actual strike of the bullet. The Army has determined that 18 rounds is the average number of rounds needed to perform zero, and we use this number for all options.

Table 6.1
Options to Reduce Rounds During M16/M4 Record-Fire Practice and Record Fire

Baseline: Semiannual—Zero = 18 Rounds, Record-Fire Practice = 40 Rounds, Record Fire = 40 rounds	Option 1: Shoot Fewer Engagements	Option 2a: Reduce Events—One Record Fire, Second if Necessary	Option 2b: EST Practice—One Record Fire, Second if Necessary	Option 3a/b: Shoot Annually, Not Semiannually
Means of reductions and amount of ammunition reduced	Reduce number of rounds per event—80 rounds	Reduce the number of events—up to 80 rounds	Reduce the number of events—up to 80 rounds	Reduce the frequency of events—98 rounds
Specific changes to current strategy	• Record-fire practice—20 • Record fire—20	• Record fire #1—40 • Only shoot record fire #2 if not qualified first round	• EST record-fire practice • Record fire #1—40 • Only shoot record fire #2 if not qualified first round	• 3a: Shoot baseline once a year with EST use semiannual • 3b: Shoot baseline once a year with no EST use
Basis for change	M16 and M1 studies that early events predict later events	• Prediction studies • Soldier experience—some will qualify without practice	TRADOC Analysis Center (TRAC) and Army Research Institute (ARI) studies	• Experience of force • EST studies that practice predicts record fire
Possible test evaluations	Collect detailed outcome data from record-fire qualification to determine if some distances or engagements can be eliminated	Survey study to determine percentage of soldiers that qualify first round	Controlled study to determine percentage of soldiers that qualify first round	Controlled study to determine degree of skill decay if shooting is annual, with and without EST use in between

and their scores shooting 20 rounds from the prone position.[6] According to Hagman, "rifle marksmanship proficiency, heretofore measured on the basis of 40 rounds, can be accurately predicted on the basis of only 20 rounds fired." Relatedly, Hagman[7] found similar predictive utility for M1 tank gunnery. Specifically, he developed predictive models that used the results of an initial set of engagements on tank Table VI[8] to predict total tank Table VI score. Hagman reasoned that, by using this predictive approach, it was possible to achieve at least a 20-percent reduction in resources used for tank gunnery.

Even though both we and Hagman would warn against making such a change without further experimental or survey-based analyses, his findings clearly indicated that shooting fewer engagements could be one approach to reducing training ammunition expenditures while still having a valid qualification event. However, the issue is that the reduction of live-fire practice may result in reduced soldier marksmanship skills, either immediately or over time.

---

[6] J. D. Hagman, *More Efficient Live-Fire Rifle Marksmanship Evaluation*, Arlington, Va.: U.S. Army Research Institute for the Behavioral and Social Sciences, Technical Report 1170, 2005.

[7] J. D. Hagman, *Enhancing the Efficiency of Tank Gunnery Evaluation: A Strategy Revisited*, Arlington, Va.: U.S. Army Research Institute for the Behavioral and Social Sciences, Technical Report 1114, 2001.

[8] In 2001, it was tank Table VIII.

Consequently, we recommend that if the Army were to pursue Option 1 in some form, additional research be performed. This research needs to include the collection of rifle marksmanship outcome data that would enable the Army to determine if such reductions in the number of live-fire engagements would reduce soldier proficiency.

If Option 1 were deemed viable, it would offer the Army an advantage over the options that reduce frequency in that record-fire qualification could still be conducted semiannually and possible skill decay between rifle qualification events could be mitigated and more frequent validation of soldier skills would be possible. The risk would be that the reduced number of engagements would lead to lowered levels of shooting skills over the long term.

Option 2a—Shoot One Record Fire and Only Fire a Second Record Fire if Required

Option 2a would achieve a reduction in training ammunition in a manner somewhat similar to Option 1. However, instead of reducing the number of engagements, Option 2a reduces ammunition by up to 80 rounds per soldier per year by eliminating the practice record fire.

Specifically, soldiers would shoot the same engagements and number of rounds (40) during record fire as they would in the current strategy. If the soldier's score after the first record fire was a qualifying score (23 or more targets hit), then the soldier's individual qualification would be complete.[9] If a soldier did not qualify with the first 40 rounds fired, then he or she would progress to the additional 40 rounds.

The basis for Option 2a is similar to that for Option 1. That is, if a subset of scores could accurately predict total score, then fewer rounds would be needed to complete record-fire qualification. To understand how many rounds could be deducted from the strategy, it should be determined what percentage of soldiers could qualify with the first 40 rounds. If this percentage is very low, then little to no savings would be realized by implementing this method. We speculate that this strategy may work better for experienced shooters.

To understand if this strategy is practical, additional predictive model studies should be completed to determine the accuracy and limitations of this and other predictive approaches to reducing the number of rounds fired during rifle record-fire practice and qualification. These additional studies should include an analysis of the differences between experienced and inexperienced shooters, as well as other common control variables, such as MOS, deployment history, amount of EST practice, and prior record-fire scores. These studies would not only indicate if predictive models could work, but also if an adequate percentage of soldiers would qualify the first time with similar scores to make this option useful.

The risk of this strategy is that it may reduce the number of more skilled shooters (sharpshooters and experts), and these numbers could continue to decrease over time. Therefore, examination of this aspect should be included in studies of this option and the other options in this chapter.

---

[9] Again, our examples throughout this chapter assume only primary iron sights are assigned to a soldier. If soldiers have additional sights, the same option procedures would be necessary with each sight, and the reductions would increase accordingly.

Option 2b—Conduct Extensive EST Practice, Followed by One Record Fire, and Only Fire a Second Record Fire if Required

This option is similar to Option 2a, except that it would incorporate structured EST training before the first record fire.[10] Basically, Option 2b would incorporate the EST as a systematic and standardized gate that would have to be completed successfully before the shooter could progress to the record fire.

The possible number of rounds reduced per soldier per year, up to 80, is similar to what could be saved with Option 2a. However, if prequalification/gate training with the EST increased the percentage of soldiers who qualified on a first record fire, then there would be further reductions in training ammunition across all soldiers with Option 2b. The additional EST practice may also result in higher average scores than Option 2a.

To determine if the EST could improve average scores and the percentage of soldiers who could qualify in the first round, additional research is necessary. Generally, early studies of the relationship between the EST and live-fire scores were inconclusive and were not specifically designed to answer the question of whether the EST improves scores. A study testing this relationship and the viability of Options 2a and 2b would require identifying shooters from EST and non-EST units and organizing them into three groups[11]: (1) no practice, record fire only; (2) EST practice, practice record fire, record fire; and (3) no EST practice, practice record fire, record fire.[12] Scores would be collected at different points during units' training cycles. Table 6.2 contains a graphical representation of this research design. The comparison of scores across all groups would enable the Army to determine what percentage of soldiers would qualify the first time with no additional practice and their average scores, as well as what percentage of soldiers with an EST gate prior to practice record fire would qualify the first time and their average scores.

Using Table 6.2 to determine whether practice has any value, Score 3 can be compared across all groups. If there are no differences across the three groups, then soldiers who did not require practice obtained the same level of proficiency on average. However, if there are differ-

Table 6.2
Study Design Template for Option 2b

Groups	Group Description	Score 1	Score 2	Score 3
One	No practice, record fire			Record fire
Two	EST practice, practice record fire, record fire	EST practice score	Practice record fire	Record fire
Three	Baseline—Practice record fire, record fire		Practice record fire	Record fire

[10] Part of the current strategy includes PMI with the EST when the simulator is available. The role of the EST in this option is different than in PMI in that it would be used as a gate similar to that used today with COFT in tank gunnery. That is, the soldier would have to demonstrate a level of marksmanship skill before being permitted to progress to live-fire training. This would require development of the tasks and standards for the EST gate, as these do not currently exist.

[11] Ideally for study design reasons, each unit would have shooters randomly assigned to one of the three groups. However, we expect that this will not be possible and that groups and units will be correlated. That is, all shooters in a unit would be in one of the three groups. If this assumption is correct, then additional control variables and larger sample sizes will be required to achieve statistically reliable and valid results.

[12] All soldiers would complete PMI as they regularly would. Soldiers in the EST would then conduct a gate-style EST gunnery event in which they would have to accomplish a predetermined skill level prior to going to the practice record fire.

ences, or there are differences for some subgroups, then the Army would know what types of practice yield a greater percentage of qualified shooters. For instance, it may be possible that the percentage of experienced shooters qualifying does not differ for Score 3 across all three groups, but a greater percentage of inexperienced shooters qualify in the EST group than the other two groups.

Again using Table 6.2, the application of this research design could enable the Army to compare the percentage of soldiers who would qualify on the first record fire. In this case, Score 2 and Score 3 values for Groups Two and Three should be compared. If the correlation between the shooters' Score 2 and Score 3 are positive, large, and statistically significant, then Options 2a and 2b may be viable.[13]

The study design in Table 6.2 also would allow for an assessment of the EST's value as a rifle marksmanship trainer. Contrasting the Score 2 values for Groups Two (EST practice) and Three (no EST practice) would indicate the value of using the EST as a gate for rifle record fire. If shooters in Group Two had better scores than those in Group Three, it would suggest the EST has value within a rifle training strategy.[14] Moreover, if Score 2 for Group Two is higher than Score 3 for Group 3, then practice with the EST (Option 2b) may be of more benefit than a practice record fire (Group 3—the baseline).

The first three options consider ways to reduce the training ammunition expended without decreasing proficiency, either by reducing the number of rounds fired per event (Option 1) or by reducing the number of live-fire events (Options 2a and 2b). However, if strategy developers do not want to reduce the rounds per event or the number of events, another set of options is possible. We next consider two alternative options designed to achieve reductions by reducing the frequency of events.

Options 3a and 3b—Conduct Current Rifle Qualification Annually, Not Semiannually

Options 3a and 3b are designed to reduce the amount of training ammunition fired by decreasing the number of times soldiers conduct rifle qualification. For the purposes of detailing these two options, we selected 12 months between rifle qualifications as opposed to the current baseline of six months. The fact that all but one of the foreign armies and the USMC have annual, rather than semiannual, rifle qualification strategies makes this a reasonable option to consider. However, any reasonable amount of time between qualifications could be selected and, just as important, tested.

In Options 3a and 3b, soldiers would shoot once, not twice, per year, making the reduction in training ammunition per soldier 98 rounds. Option 3a would include a full EST training event, including qualification, in between the annual record-fire qualification events.[15] Option 3b is set so that rifle qualification occurs annually with no "makeup" event in between. The advantages of both options are that they would decrease the amount of training ammunition and that no changes to the current rifle record-fire table would be necessary.

---

[13] Also, it would be important to test for score transposition. That is, are Score 2 values greater than Score 3 values? If so, using the first record-fire scores to qualify could have some risk.

[14] Even if there were no differences found for Score 2, we recommend comparing Score 3 values across the three groups because the value of the EST might be realized after record-fire practice.

[15] Other EST usage alternatives are possible. For example, monthly or quarterly structured practice events. However, for this report, we assumed that the frequency of marksmanship events would not increase and that soldiers would qualify/conduct individual marksmanship every six months or less.

Yet both options share two possibly serious disadvantages. First, there is an increased risk of skill decay between record-fire training events as compared to the baseline. Second, leaders will have fewer opportunities to validate the rifle marksmanship skills of soldiers with a live-fire event.

Similar to Options 1, 2a, and 2b, it would be both practical and informative to conduct research that examines skill decay when the frequency and number of events and live-fire engagements are reduced. In fact, we argue such testing is necessary to give reasonable assurances that there would not be a loss in shooting proficiency with this or any revised strategy.[16] We found limited research specifically related to the decay of rifle marksmanship skills. Such studies are somewhat difficult to carry out because of issues connected with stabilizing sufficiently sized groups of shooters and because the factors that may influence marksmanship scores over a year or more may change.[17] However, the benefit of this type of further research could outweigh the time and resources required to execute it.[18] We next outline two possible research approaches.

The first approach would involve gathering survey data from individual shooters and their record fire practice and scores from current or very recent qualifications. Two key survey variables to collect are time since last record-fire qualification and last qualification scores. These variables can then be compared with current record-fire scores. The analysis would therefore compare changes in scores between record-fire qualifications as a function of time between qualification events. This approach will only provide reliable and valid results if (1) shooters can accurately recall their previous qualification date and score and (2) there is enough variation across shooters in time between qualification events. The first of these is straightforward; that is, if the variables are not reliable, the results are not valid.

The second point may be less obvious. Basically, to test for differences in outcome scores attributable to the time between events, there needs to be a range of times between events across all shooters.

There are other challenges with this first research approach in that it may not test skill decay, which is possible in both Options 3a and 3b, because there may not be a sufficient number of soldiers who used the EST between rifle qualification events. Even if there were sufficient numbers of such shooters, it is very likely that there would be great variation in the way they used the EST and conducted other activities that may influence marksmanship skills, such as PMI. So this research study approach has limitations. The second research approach solves some of the challenges of this approach.

The second research approach is a quasi-experimental design that would enable the Army to determine if there is skill decay with (Option 3a) or without (Option 3b) the use of the EST between qualification events, as well as the magnitude of decay. In this research design, before record fire qualification, soldiers would be selected and assigned to one of six experimental cat-

---

[16] Quantitative analyses cannot address the issue of validation for leaders. However, if the Army was confident that skill decay was no greater for the 12-month rifle qualification option, then leaders' concerns could be mitigated as well.

[17] One study (see Eva L. Baker, *Analysis of Rifle Marksmanship Data from the Marine Corps Central Master File*, Los Angeles Calif.: CRESST/University of California, November 2006) examined the rifle marksmanship scores in the USMC personnel database over a five-year period and looked at the effect of different time gaps between qualification events. The results indicated that there was limited effect, but missing data limited the conclusions that could be drawn.

[18] We say this because the study approaches outlined below require no additional rounds beyond those already allocated for unit rifle training, so there would only be small additional costs.

egories presented in Table 6.3. All soldiers would conduct baseline record fire at time "0," and their practice and record-fire scores would be recorded. Groups 1 and 2 would then re-qualify six months after time 0, Groups 3 and 4 would re-qualify after nine months, and Groups 5 and 6 would re-qualify 12 months after time 0. The odd-numbered groups (1, 3, and 5) would not use the EST between events, and the even-numbered groups would conduct an EST event similar to record-fire qualification between time 0 and the second record-fire qualification.

The analysis of this design would contrast the changes in qualification scores across the six groups and would provide the basis to answer two questions. For example, contrasting the changes in scores between the odd and even groups answers the following question: Does EST training mitigate skill decay between events?[19] Comparisons within the odd-numbered groups answer this question: Does the rate of skill decay increase as the time between events increases for soldiers who do not use the EST? The question of whether skill decay differs as a function of time between events and EST usage is answered by simultaneously contrasting all of the groups.

Combining Training Ammunition Reduction Approaches

Thus far in this chapter, we have detailed options that separately consider three primary modes of training ammunition reduction: (1) reduce the number of rounds per event, (2) reduce the number of events, or (3) reduce the frequency of the events. In this section, we consider a combination of these reduction modes. We recognize that there are a large number of possible combinations; because there are so many, we present only one as an illustration. Also, we are not suggesting that it is the best way to change rifle training strategies. Instead we offer it as an illustration of a possible way ahead.

Figure 6.1 graphically depicts this example case. This case includes the use of the EST as a practice and training tool, the use of the EST as a gate, only conducting one record fire (when appropriate) and a possible reduction in the number of targets per event. Moving from left to right in Figure 6.1, the shooter first conducts record-fire prequalification with the EST. If the shooter's EST score is a qualifying one, the shooter moves on to record fire. If not, the shooter conducts detailed, paced marksmanship training using the EST under the tutelage of an expert trainer/shooter. After successfully completing EST prequalification, the shooter conducts an "initial" live record-fire event engaging 20 targets.[20] If the score on the initial 20 targets is a qualifying one, the soldier is considered qualified.[21] If the soldier does not qualify,

Table 6.3
Skill Decay Research Design Template for Options 3a and 3b

Time Between Qualification	No EST	EST Between Qualification
Six months	Group 1	Group 2
Nine months	Group 3	Group 4
Twelve months	Group 5	Group 6

[19] Multiple comparisons are possible using this research design, and we do not detail each here.

[20] Our selection of 20 targets instead of 40 is somewhat arbitrary. The actual number of targets should be based on prediction studies similar to those conducted by Hagman (2001) or as we illustrated in Table 6.2.

[21] There are three levels of gunnery qualification: expert, sharpshooter, and marksman. These levels should remain with any revised strategy. But for ease of presentation in this report, we simply consider a shooter qualified or not qualified.

Figure 6.1
An Example Combining Training Ammunition Reduction Approaches

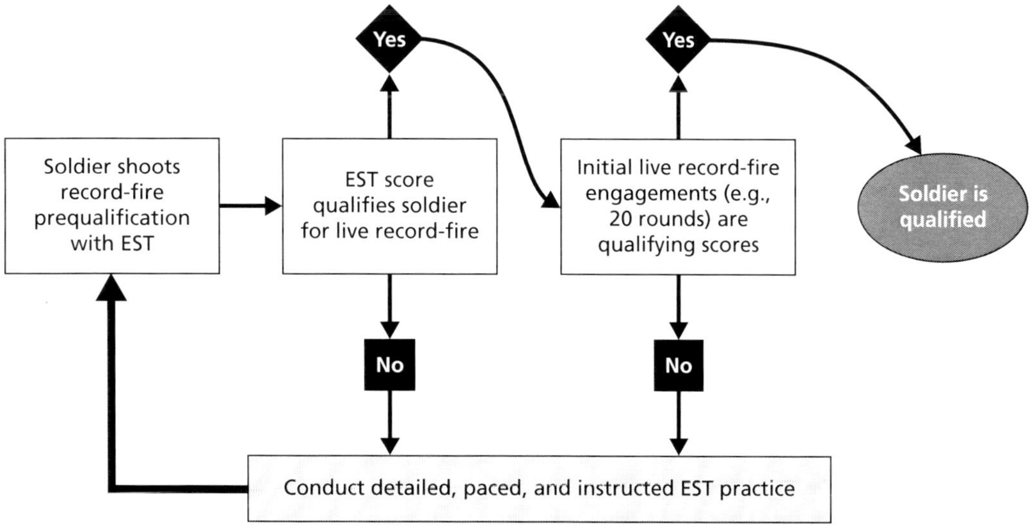

RAND *RR448-6.1*

he or she returns to conduct additional EST training, the EST gate, and again the live record fire. This process continues until the soldier is qualified.

The process in Figure 6.1 could be conducted at different points in time based on the soldier's experience and prior qualification scores. For example, an experienced soldier who hits 20 out of 20 targets on the initial practice record fire may not need to complete record-fire qualification for another 12 months. On the other hand, soldiers who needed several iterations and still hit a low number of targets may need individual live fire every six or nine months.

No matter what changes to record-fire qualification strategies are considered, an undesirable outcome is a decline in the marksmanship of U.S. Army soldiers. To prevent this decline, good research is needed to support sound decisions. The prior research that we found and documented in this chapter provides some preliminary insights (e.g., the predictive models developed by Hagman[22]); but more research is needed. The combinatorial approach presented in Figure 6.1, or other similar approaches, should only be considered if the following three questions can be answered:

1. Can a subset practice record-fire score reliably predict total score?
2. Does using the EST as a practice aid or as a component in a gated strategy correlate with live record fire?
3. Will live-fire scores decline as the time between qualification events increases, and can EST training mitigate the decline?

---

[22] Hagman, 2001; Hagman, 2005.

Conclusions and Implications

Based on the limited research to date, as well as our own observations, it seems possible to reduce the training ammunition in the rifle record-fire qualification strategies, and a better mix of live-fire and virtual capabilities has the potential to support such reductions. But some well-done studies along the lines of those described in this chapter will be necessary to guard against changes that result in lowered soldier skills in this important area.

A major caveat we would add is that we are discussing potential changes to the current rifle strategy outlined in DA Pamphlet 350-38. As discussed in Chapter Two, the degree to which units follow the strategy is not clear, and training ammunition expenditure and EST usage rates indicate they do not. So modification of the strategy will not necessarily save ammunition.

A second, and more significant, caveat is that in this chapter we looked at greater efficiency for only a portion of current rifle training strategies—record-fire qualification. The issue that many important combat skills are not being trained and validated is arguably of equal or greater importance. Also, the issue that the Army's rifle training strategies do not include allocation of ammunition to events such as grouping exercises, which improve basic marksmanship skills before record fire, needs to be considered. These two issues mean that a case can be made that some or all of any possible savings in record-fire qualification should be wholly or partially reinvested in rifle training strategy improvement rather than efficiency.

An implication of this examination of options for making rifle training strategies more efficient is that it may well be possible to improve weapon training strategies, and that changing the mix of live and virtual weapon training modalities may be a reasonable approach to making improvements. However, it is important to study any option before implementation to ensure it results in savings without reducing unit weapon-proficiency levels. A range of study options is available, including those presented in this chapter. Some types of studies can be performed faster than others, but controlled studies that are carefully designed and executed provide a stronger basis for making decisions. The major point is that certain types of studies are beneficial because they will provide commanders with better information for shaping their unit training programs and provide the Army's leadership with data that can allow them to make more informed risk assessments when considering weapon training strategy changes.

Another implication is that establishing a system to gain a better understanding of the true nature and effectiveness of unit rifle training programs would benefit the process of developing and assessing options for improving weapon training strategies. For example, if a large number of units are already using the strategy of having soldiers zero and go directly to record-fire qualification, with only those soldiers who do not qualify initially refiring the record-fire table, then the actual training ammunition savings will be limited.[23]

---

[23] Survey data collected by the Maneuver COE indicate that this is the case in many units.

CHAPTER SEVEN

Improving the Wider Range of Weapon Training Strategies—Conclusions, Directions, and Implications

In this chapter, we first present conclusions we drew from applying the results of our examination of small-arms training strategies to the full range of weapon training strategies outlined in Chapter Two. Second, we outline directions the Army could take to improve its processes for developing and improving its weapon training strategies. Finally, we present the implications this examination of weapon training strategies has for the broader range of the Army's training and leader development programs.

Conclusions

Training Strategies and the Resources Needed to Support Them Should Vary by Weapon System

Our review of small-arms training strategies showed that each weapon system is different, requiring different sets of tasks and skills; therefore the training strategy for each needs to be shaped and supported differently. For example, small arms are somewhat unusual in that accurate engagement is mainly a function of individual psycho-motor skills. Other major weapon systems are more supported by technology and require crews or units collectively applying procedures rapidly and precisely. Even within the category of small arms, there are differences between machine guns and rifles. For example, the rifle is mainly a point-fire weapon with individually aimed shots. The machine gun is fired in bursts, can be "walked" onto point targets, and has a major area fire role that is more difficult and expensive to exercise in live-fire events than point-fire engagements for the rifle.

There Is Potential to Improve the Effectiveness and Efficiency of Weapon Training Strategies by Changing the Mix of Live-Fire and Virtual Modalities, but Improvement Will Require Effort and Study

Supporting TRADOC's efforts to examine changing the mix of live and virtual modalities to improve weapon training effectiveness and efficiency was a primary motivator of this report. Our conclusion is that simulators have the potential to improve weapon training strategies and that this potential is likely to increase as weapon system technology and simulator capabilities improve. However, while there is this potential, determining the specifics will be a challenge because each weapon system is different, each requires its own examination, and the examination needed to determine that changes do not unduly risk undercutting engagement proficiency could be extensive. The consideration that the staffing resources the proponents of weapon system training have for such an effort are limited further complicates this issue.

Also, while the increased use of simulators seems an attractive concept for supporting improved training outcomes and reducing costs, it will need testing and study to ensure that training improvement can be achieved. For example, the Army would need to verify that training "moving target" engagement on the EST develops the live-fire skills needed to hit a moving target. If not, then either EST training on this type of engagement should not be a part of the strategy, or the EST's capability to train this skill should be improved.

A second consideration is that the ability to take greater advantage of the use of simulators will likely require additional investment. We found that the number of EST training sets was not sufficient to support both preliminary marksmanship and the qualification requirements in current small-arms strategies. To support training additional advanced marksmanship skills, more capacity would be needed. Thus, we found that realizing significant training ammunition savings while achieving critical standards would likely be possible only if the Army invests in a low-cost weapon skills simulator and institutes formal programs to train unit small-arms trainers. Likewise, as outlined above, the ability of the simulator to train needs to be verified and, if there are shortfalls, investments for improvement would be needed.

Moreover, our examination of a broad range of weapon training strategies showed no obvious examples of where significant savings would likely arise from changing the virtual-live mix. Simulators already play a large role in many weapon training strategies—especially ones with high costs per round.

The most feasible targets would be weapon systems for which the fielded simulator capacity is not fully used, since fielding added simulator capabilities would require up-front procurement and development resources before any live-fire savings or weapon training improvements could be realized. Investing in additional simulator capabilities before there is a demonstrated possibility of major savings or enhancement would be difficult in light of ongoing budget cuts.

But before assuming that an underutilized simulator system represents a potential for expansion, the Army should determine the reason for the underutilization, especially when operational deployments may have suppressed recent demand. A problematic issue would arise if unit leaders did not see increased usage as being of benefit, in which case increasing simulator use in strategies will have little likelihood of achieving greater usage and improving strategy outcomes.

Thus our examination supports the conclusion that changing the mix of virtual and live training in current weapon strategies will require a structured effort to achieve actual improvement. Decisions on leveraging simulator technology require close examination on a case-by-case basis. In each case, the capabilities of the simulator technology and a structured total costs/needed training outcome approach would be needed. Further, the examination should be supported by such types of tests and studies as outlined in Chapter Six and done as part of a review of the weapon's overall training strategy. Key factors to consider include the following:

- critical combat engagement tasks skills
- the cost of simulator procurement, operation, sustainment, and unit/solder time
- the cost of live-training ammunition, fuel and repair parts, range support and maintenance, and unit/soldier time
- the capability of simulator training—that is, the tasks/skills trained and the effectiveness of training
- the capability of live-fire training—that is, the tasks/skills trained and the effectiveness of training.

Identifying and Verifying Combat-Critical Weapon Engagement Tasks and Skills Is an Important First Step to Both Developing and Refining Weapon Training Strategies

The most important finding from our examination of small-arms strategies is that many critical small-arms tasks and skills are either not addressed at all or not adequately addressed in that they are not qualification requirements. An example of the former is area fire for machine guns, while an example of the latter is area short-range reflex rifle engagements.

This means an important conceptual starting point for improving weapon training strategies and determining the level of resources appropriate for their support is using combat requirements as a basis for determining the tasks, skills, and standards that are critical to combat success. Figure 7.1 is a simplified, conceptual outline of what such a process would entail. Note that such a process would be continuous.

Weapon training standards should be based on the weapon system's operational requirements developed by the combat development community. This would involve considerable qualitative military judgment and include determining what types of engagements and what levels of engagement accuracy and speed are needed for combat success. While defining specific unit training objectives is the responsibility of the chain of command, determining base weapon-training engagement standards (i.e., the minimum level of engagement skills needed across the Army) is not one that should be decentralized.

A key point is that, while weapon training strategies have changed little since 2000, operational requirements and the capabilities of many weapon systems have. In the case of small arms, basic combat requirements have not changed, but weapon capabilities (e.g., range and optics) have. In the case of other weapon systems, the nature of threats may affect the types of engagements that are critical to combat success (and their standards). For example, to avoid collateral damage, artillery engagements may need a greater emphasis on clearance and accuracy and less emphasis on response speed. Tank engagements may need to place more emphasis on machine gun area fires and less on the engagement of armored vehicles. Further air-defense fires against unmanned aerial vehicles may become a critical task. We are not suggesting these as specifically needed changes, but rather as the type of changes that training developers should be considering. A major consideration is that the resources available for weapon training are finite, and it will not be possible to achieve desirable levels of proficiency for every task and

Figure 7.1
Conceptual Weapon Training Strategy Development and Revision Process

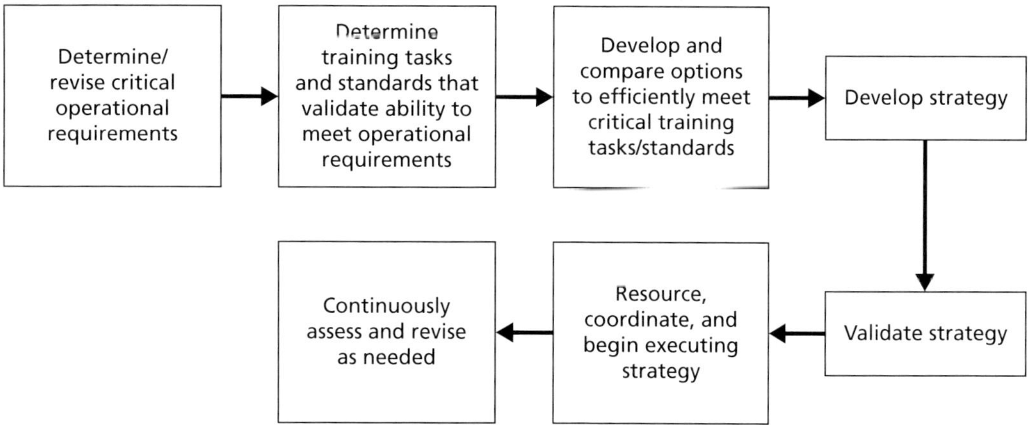

RAND *RR448-7.1*

skill. Difficult decisions will be needed to specify qualification standards that meet truly critical operational requirements within reasonably available resource levels.

The likelihood of reductions in training resources is high and, in a very practical sense, the training strategies and standards that these resources support may have to change in response. These changes must be made with a clear view toward meeting the critical operational readiness levels, so defining critical engagement skills is key to making such decisions. They also can be used to make the case for needed resources by clearly showing the risks that may result from not providing the resources for needed training.

Having a set of accepted, objective, combat-critical training standards to use as criteria for assessing alternative strategies is also important. An example is examining the potential for reducing the number of rounds in Paladin artillery training strategies. From many perspectives, it may seem logical to propose that the 160-round training strategy for this system could be reduced.[1] There are simulators for training forward observer skills and a simulator/stimulation system for training Fire Direction Center (FDC) tasks. Firing data are automatically transferred from the FDC to the guns and, unlike for towed artillery, the level of crew skills needed to orient the guns is limited. Moreover, for both Paladin and towed howitzer crews, the task of preparing ammunition for firing and other activities needed for firing can be practiced and evaluated without actually firing the round. These considerations could be used to support an argument that a smaller number of rounds could allow sufficient training for fire-mission success.

However, the field artillery training developers with whom we spoke uniformly expressed a strong belief that training the three field artillery components separately would not be as effective as training all three together, and that live-fire training was necessary to achieve needed proficiency. We could not determine the degree to which this was true or if improvements in the observer training simulation, FDC training simulator/stimulation, or the weapon system itself could change this need. But the point is that the degree to which virtual technologies can reduce the need for live-fire training is dependent on the degree to which these technologies can fully and faithfully train critical live-fire tasks and skills. If they cannot, the reasons should be identified and either the simulator technology should be improved, a decline in proficiency accepted, or the live-fire allocations should be retained. For instance, is forward observer training the problem? If so, then the next question is whether it could be improved at an acceptable cost. Moreover, not only must the technology be able to support training, but also, for adoption, the operational community must be convinced it will do so.

Finally, the objective that generated this research effort was to "evaluate the appropriate mix of live-fire and virtual training simulators to supplement or replace live-fire individual tasks resulting in an overarching Army training strategy." An objective evaluation of the overarching strategy can only be done in the context of the combat-critical outcomes these individual strategies and the live-virtual mix supporting them are to achieve.

Better Alignment with ARFORGEN Readiness Goals and Manning Strategies Could Achieve Needed Readiness Levels More Efficiently

Most weapon training strategies involve semiannual training and qualification through collective live-fire exercises. Our examination of small-arms strategies showed that better aligning

[1] Indeed, during discussions with non-artillery training staff members, many brought up the idea that 160 rounds was more than really needed to train field artillery units.

small-arms training strategies with ARFORGEN processes could promote increased readiness levels and the more efficient use of weapon training resources. The realignment could occur by shifting from repetitive and cyclic to progressive weapon training strategies that build to full readiness at the time of entry into the ARFORGEN's Available pool. The benefits of using this general approach could yield similar benefits for many other weapon systems as well. While the specifics of events and their timing would differ, the general concept would be to have preparation activities in the first half of the Train-Ready pool phase, individual and crew qualification toward the middle, and collective live-fire and designated crew qualification toward the end. Such a concept could lead to a more efficient program peaking to support full, sustained Available pool readiness. Adopting this concept would essentially move weapon training strategies from a semiannual to annual basis. The number of qualification events could be reduced, better integration with ARFORGEN manning strategies could be achieved, and the needed Available pool readiness level could still be attained.[2]

The Lack of Data Undercuts the Army's Ability to Make Needed Improvements

Assessing alternative weapon training strategies is complicated by the fact that the Army does not have a system for knowing how closely units follow current weapon training strategies, the degree to which they meet standards, or how well current standards relate to combat success. In our examination of small-arms strategies, the question of how well soldiers can actually shoot was raised. Some thought that, because of the extent of operational experience, the emphasis on small-arms proficiency in predeployment training programs, and the tactical success in Iraq and Afghanistan, the current programs have been validated and there is no practical need for improvement. Others thought the skills were generally very low, that many unit weapon training programs were minimal, and that tactical success was often more a result of supporting fires and the volume of fire than accurate small-arms fire. However, no systematic data are available to answer this question and support Maneuver COE training development efforts to improve small-arms training strategies.[3]

Coordinating Weapon Training Strategy Changes to Ensure Needed Training Resources Is Important

Conducting weapon training requires ammunition, ranges and targetry, simulations, and other types of resources that come from many different Army PEGs, programs, and program subcomponents. This means that coordination for the balanced set of resources needed to support weapon training strategies is complex and requires integrating the efforts and priorities of many different organizations.

An important area where integration is needed is training ammunition, which is provided from the Equipping PEG. This arrangement especially complicates changing the mix of live-fire and virtual modalities in weapon training strategies. If an increase in simulator numbers or capabilities is needed to implement the changed mix, then up-front resourcing will be required

---

[2] Note that this direction would not necessarily reduce the amount of weapon training or the amount of rounds for live-fire; it reduces the time and number of rounds allocated for qualification events, and by doing this, increased time and rounds could be made available for structured preparation training. However, as with any major initiative, a formal assessment program should be established and modifications made as analysis shows the need for change.

[3] An online survey conducted by the Maneuver COE indicated a wide degree of variance in rifle marksmanship training program across units.

before training ammunition savings can be achieved. The Vice Chief of Staff of the Army must approve such a shift. While this is not a major barrier, it does complicate the ability to make such changes.

Because resources are limited and demands are many, an important consideration is ensuring that weapon training strategies are feasible and supported. For example, collective live-fire exercise requirements should align with available range capabilities. Also, as discussed above, achieving the best possible balance of simulator and live-fire capabilities would facilitate increased outcomes and efficiency. This means that weapon training strategies must be developed considering resource limitations and what is feasible in terms of overall training strategy needs.

More Definition of Processes for Weapon Training Strategy Development and Review Would Enhance Improvement

A conclusion presented in Chapter Two was that current weapon training development processes focus on incremental improvement, generally assuming that current strategies only need slight adjustments. However, our review of changes in training requirements and the training environment and what these mean to small-arms strategies indicates that some major changes are needed to meet critical combat engagement requirements. We also identified directions that could be taken to make these strategies more efficient.

Our general review of weapon system training strategy development processes and more specific review of small-arms strategy development lead to the conclusion that the wider range of weapon training strategies could benefit from a similar review. Even if the further review of other strategies does not uncover a need for change, the validation that such a comprehensive review would provide would be valuable. The Army will be dealing with constrained budgets for the foreseeable future, and it will be looking for savings. Training ammunition budgets are under scrutiny now, and this oversight will undoubtedly continue. The reality is that there are probably areas where training ammunition savings are possible, and it is important to be proactive in objectively identifying areas where reductions can reasonably be taken and where substituting virtual for live-fire training is possible. In cases where there are risks, it would also be important to identify the possible readiness effects. Targeted reductions and strategy changes that preserve needed readiness levels are far preferable to potentially more damaging "across the board" cuts.

While the function of performing periodic comprehensive weapon system training strategy reviews, including the best possible integration of live and virtual capabilities, is a logically required responsibility of each weapon system proponent, there is no specified requirement for such a review in DA Pamphlet 350-38; Army Regulation 5-13, *Total Army Munitions Requirements Process and Prioritization System*; or TRADOC Pamphlet 350-70-1, *Training Development in Support of the Operational Domain.*

A more basic issue is that the Army has not defined an analytical process for developing weapon training standards and strategies or for integrating them with other training strategy components. Of special concern is the lack of a requirement for a development and review process to support the better integration of training ammunition and range programs with simulator programs.

Developing the capability for comprehensive development, review, and improvement processes will be a challenge. A basic issue is that, because there is no directed requirement, proponent staffs are not designed and manned to support the comprehensive development, continuous review, integration, or change of weapon system strategy.

Directions for Improving Weapon Training Strategies

Based on these conclusions, the Army should consider three major directions to improve weapon training strategies. These include additional guidance, data collection and analysis, and strategy improvement as part of an integrated process.

Develop Definitive Guidance on Weapon Training Development Processes

The Army should develop guidance on how weapon training strategies and standards are developed and more clearly identify the requirements and responsibilities for developing weapon training strategies. This guidance should be included in Army Regulation 5-13, DA Pamphlet 350-38, and TRADOC Regulation 350-70-1. The guidance should outline a logical, integrated process for developing and refining weapon training strategies and standards that efficiently meet critical operational needs. The process illustrated on Figure 7.1 could serve as a general model for such a process. Key elements in the guidance should include the following requirements:

- the identification of critical engagement tasks and standards based on the weapon system's required operational capabilities
- the development of weapon training strategies to achieve critical tasks and standards at the lowest total costs possible and the use of the best possible mix of training modalities to achieve them
- the validation of weapon training standards before implementation
- the coordination of weapon training strategies across the full range of training strategies and programs
- a continuous process for reviewing, assessing, and improving the strategies, including monitoring unit implementation, tracking qualification rates and resource utilization, and identifying areas for improvement
- the identification by weapon system proponents of where virtual simulators could enhance or make the strategies more efficient and initiate the development of the requirement documents needed to support procurement and fielding.

Collect and Analyze Data on Unit Weapon Training Programs and Operational Weapon Proficiency Levels

Weapon system training developers have several data sources to draw from. Unit Status Reports provide data on the number of crew/squads that are manned and qualified, TAMIS provides data on overall unit expenditures of training ammunition, and there are simulator usage reports. But these sources lack many key data elements needed to improve weapon training strategies and define and defend the resources needed to support those strategies. For example, for small arms, the infantry-squad qualification data do not show how many soldiers qualified or at what levels (e.g., marksman or expert).

Also there are issues concerning the degree to which the strategies are actually executed. The DA Pamphlet 350-38 requirements for an infantry soldier to be qualified on a rifle include unassisted night fire and CBRN qualification on the EST. Yet EST usage rates are low, and it is clear that many soldiers do not qualify on these exercises. What is not clear is if they are doing these events in live fire or simply not adhering to the qualification requirement.

TAMIS data can show how many rounds a unit uses per year, but they do not show the events that were supported. Nor do they show how the rounds were used, such as how many rounds were used for re-fires for soldiers and crews who did not qualify the first time, or how many rounds were used on the preparatory tables, collective live-fire exercises, or other commander-chosen live-fire events. There are similar issues for simulator usage reports.

So overall these data are general, not always accurate, and located in different databases. The result of these factors is that these data are of limited practical use to weapon-system training developers.

Additionally, each proponent makes efforts to get feedback from field commanders, but, as we have argued, they are not able to do this in a systematic, comprehensive way. This is not to say that proponent training development staffs do not have some direct knowledge of unit programs. They and their supervisory chain have members with extensive unit experience, including military personnel with recent unit and operational experience, but this experience can be dated or based on experience in a small number of units.

The result is that, as a group, training developers do not understand many key factors needed to adapt and defend current strategies. Some questions are key. For example, what types of weapon training are actually being done, and which elements of the doctrinal strategies are not being followed? How is training ammunition being used? Why are EST usage rates low? How well are soldiers shooting in combat, and is there a need for improvement? In a period in which there is a need to reevaluate and revise training strategies to meet a different balance of operational requirements and during which there will be great pressure to reduce training budgets, the answers to such questions will likely become even more important for the weapon training community.

Develop and Improve Weapon Training Strategies as Part of a Larger, Integrated Process

Efforts to develop and improve weapon training strategies should be accomplished as a part of an overall systematic training strategy and resource data collection and analysis process. The Army's analytical resources to support training development are limited. Developing the capability to collect unit training-program data to support development will be a special challenge.

The development of weapon training strategies without considering or integrating these with other training requirements (e.g., mission-command training, maneuver training, and leader development) makes little sense, especially when changes in one area affect others and units have a common set of resources (primarily budget dollars and unit time). Likewise, collecting data on unit training activities, the resources used to support them, and areas for improvement would be more efficient if done as a coordinated effort. This would also provide a more integrated examination across training strategy elements, especially considering that the major constraint on unit activities is unit time.

Revised training strategies should aim at defining a feasible balance of training activities needed to meet operational requirements. The core reason the Army has a program to manage weapon training is to determine the needed weapon-training resourcing levels and make sure commanders have enough resources to achieve mission-ready training levels before deployment. This rationale applies to the complete range of the Army's training and leader development programs. The Army cannot afford to buy and build all the training capabilities needed to give commanders full flexibility to shape their training programs. Rather, it must make these resourcing decisions from the perspective of providing as balanced and complete a set of truly needed training capabilities as possible.

This means difficult decisions must be made. Given changed operational requirements, this logically suggests that some capabilities should grow and others could be reduced or eliminated. For example, it might well be that the importance of small-arms engagement has increased. If so, the difficult decisions for the Army include determining what changes are needed and where increased resources to support any needed changes are to come from.

Implications

The findings and conclusions of the weapon training research supports and reinforces the findings, conclusions, and suggested directions presented in a recently published RAND Arroyo report examining the broader range of Army Training and Leader Development (ATLD) programs.[4]

The weapon-training strategy research documented in this report examines a different training activity and its management processes in depth. However, this research resulted in the same overall conclusions and identifies the same areas for improvement. The overall conclusion of the earlier report is that current ATLD management processes were developed for incremental change, but now fundamental changes are likely required. There are three major process areas for improvement:

- more direct understanding and focus on operational force needs
- increased integration across strategies, ATLD programs, and other PEGs
- the development of a more structured cost-benefit approach to making ATLD program decisions.

Two important improvement areas that the weapon training research amplifies are the need to ensure that training tasks and standards address all the critical tasks and skills required for operational success and that a range of training modalities and activities will often be required to accomplish this goal.

The earlier research also outlined a number of specific directions the Army could take to make improvements. The weapon training research directly supported some of these directions. Our examination of current processes for developing and supporting weapon system training strategies strongly reinforced the potential benefits of forming a single organization

---

[4] This research is documented in James C. Crowley, Michael G. Shanley, Jeff Rothenberg, and Jerry M. Sollinger, *Adapting the Army's Training and Leader Development Programs for Future Challenges*, Santa Monica, Calif.: RAND Corporation, TR-1236-A, 2012, and was conducted for DA Director of Training and the U.S. Army Combined Arms Center's Deputy Commander-Training. The study's primary objective was to identify directions that the Army can take to improve its training and leader development management processes and architectures. The directions for improvement outlined in this report were developed from an examination of four important training activities: (1) the Army's Distributed Training Program, (2) Brigade Combat Training Team training strategies, (3) the Advanced Leaders Course (an important part of the Army's NCO Professional Military Education program), and (4) the Army's Training Support System (TSS) management processes that provide products, services, and facilities to support ATLD strategies.

with the overall responsibility for data collection, analysis, and the support of DA ATLD program management processes for integrating the overall effort.[5] Currently, many organizations are working on different parts of this effort, but none is doing overall data collection and analysis to support it. Two examples in the weapon training area are the collection of training and operational effectiveness data and the integration of training ammunition, range, and targetry programs with simulator programs.

The overall issue for the Army is that difficult decisions must be made to shape, support, and balance the range of ATLD strategy components to best meet new requirements. The research effort documented in this report focuses on weapon system training. It shows that decisions on the amount of resources that should be allocated to weapon training should be made in the context of the best possible use of limited resources to get the best possible overall training readiness outcomes. This research reinforced the earlier report's conclusion that there is a need to provide the Army's leadership with better data and analysis to support such ATLD strategy decisions. Establishing a single organization responsible for this function is a logical, and possibly necessary, direction for the Army to take to do this. A key element would be forming this organization from existing staff resources, as obtaining additional resources for a new organization is not a reasonable expectation.

A broader implication from the earlier report that is amplified here is the need for better integration among the Training and other PEGs. For example, there are major impacts on the number of times a unit must conduct weapon qualifications during Available periods to maintain readiness if there are high personnel-turnover rates.[6] The prioritization of unit stability in manning strategies could reduce the number of times units need to conduct qualifications, which would result in reduced requirements for weapon training resources.

---

[5] The other directions documented in the ATLD report are: (1) improve the overall analytical and data collection process; (2) improve systems for collecting data on ATLD achievements, nature, and needs; (3) create improved mechanisms for managing by direct ATLD activity; (4) enhance ATLD and Army-wide information technology architectures to improve data collection and analysis; and (5) evolve emerging ATLD governance structures to improve focus on operational-force readiness. Full discussions are in the report.

[6] This is illustrated by the frequency of tank/BFV gunnery events conducted by heavy BCTs during the 2001–2002 time frame. While crew qualification was only required semiannually, these units averaged 2.6 crew qualification events per year. This caused the need to conduct "makeup" gunnery periods to qualify newly formed crews. See Shanley et al., *Supporting Training Strategies for Brigade Combat Teams Using Future Combat System (FCS) Technologies,* Santa Monica, Calif.: RAND Corporation, MG-538-A, 2007.

References

Australian Army, Defense Instructions (Army), *Army Small Arms Weapon Proficiency Policy*, Admin 89-1, December 2008.

Baker, Eva L., *Analysis of Rifle Marksmanship Data from the Marine Corps Central Master File*, Los Angeles, Calif.: CRESST/University of California, November 2006.

British Army, *Army Operational Shooting Policy*, Army Code Number 71810, February 2011.

Corcoran, Patrick E., *Training Effectiveness Analysis, Simulation and Analysis of the Training Effectiveness Analysis-TOW (TOW TEA) Flight Data*, U.S. Army Materiel Systems Analysis Activity (AMSAA), April 1980.

Crowley, James C., Michael G. Shanley, Jeff Rothenberg, and Jerry M. Sollinger, *Adapting the Army's Training and Leader Development Programs for Future Challenges*, Santa Monica, Calif.: RAND Corporation, TR-1236-A, September 2012.

Dyer, Jean L., Peter S. Schaefer, Martin L. Bink, David R. James, Richard L. Wampler, and Michael D. Dlubac, *Soldier Performance on a New Marksmanship Course of Fire*, Alexandria, Va.: U.S. Army Research Institute, Research Report 1924, June 2010.

Evans, Kenneth L., Jean L. Dyer, and Joseph Hagman, *Shooting Straight—20 Years of Rifle Marksmanship Research*, Alexandria, Va.: U.S. Army Research Institute, ARI Special Report 44, October 2000.

Hagman, J. D., *Enhancing the Efficiency of Tank Gunnery Evaluation: A Strategy Revisited*, Arlington, Va.: U.S. Army Research Institute for the Behavioral and Social Sciences, Technical Report 1114, 2001.

Hagman, J. D., *More Efficient Live-Fire Rifle Marksmanship Evaluation*, Arlington, Va.: U.S. Army Research Institute for the Behavioral and Social Sciences, Technical Report 1170, 2005.

Hagman, Joseph D., "Using the Engagement Skills Trainer to Predict Rifle Marksmanship Performance," *Military Psychology*, No. 10, Vol. 4, 1998, pp. 215–224.

Hagman, Joseph D., and John E. Morrison, *A Strategy for Effective Device-Based Tank Gunnery Training in the Army National Guard*, Alexandria, Va.: U.S. Army Research Institute, ARI Research Report 1694, June 1996.

Hughes, Charles R., Walter B. Butler, Bruce S. Sterling, and Arvid W. Berglund, *M-1 Unit Conduct of Fire Trainer-Post Fielding Training Effectiveness Analysis*, Alexandria, Va.: U.S. Army Research Institute, TRAC-WSMR TEA-16-87, ACN 67571, April 1987.

Hughes, Charles R., Maryann Morales-Steigely, and Max L. Musser, *M2/M3 Unit Conduct of Fire Trainer; Post-Fielding Training and Effectiveness Analysis*, Alexandria, Va.: U.S. Army Research Institute, TRAC WSMR TEA-90-016 CAN 023378, August 1990.

Hughes, Charles, and Kathy L. Nau, *Engagement Skills Trainer (EST) 2000, Training Effectiveness Analysis (TEA)*, Fort Monroe, Va.: TRADOC Analysis Center, March 2008.

Kramer, Ronald E., and Wade C. Rowatt, *A Review and Annotated Bibliography of Armor Gunnery Training Devices*, Alexandria, Va.: U.S. Army Research Institute, ARI Research Report 1652, November 1993.

Miller, Claude, and Michelle Zeisset, *Engagement Skills Trainer (EST) 2000, Post Fielding Training Effectiveness Analysis (PFTEA)*, Fort Monroe, Va.: TRADOC, WSMR-PFTEA-02-025, September 2002.

Murphy, Kevin, James L. Farr, and Greg Loviscky, *Study to Quantify the Benefits and Costs of Simulated Versus Live-Fire Training at USMC Ranges*, University Park, Penn.: Penn State University, January 2007.

Sabol, Mark A., John A. Ellis, and Robert A. Wisher, *Staying Sharp: Retention of Military Knowledge and Skills*, Alexandria, Va.: U.S. Army Research Institute, ARI Special Report 39, May 1999.

Shanley, Michael G., James C. Crowley, Matthew W. Lewis, Ralph Masi, Kristin J. Leuschner, Susan G. Straus, and Jeffrey Angers, *Supporting Training Strategies for Brigade Combat Teams Using Future Combat Systems, (FCS) Technologies*, Santa Monica, Calif.: RAND Corporation, MG-538-A, 2007.

Sterling, Bruce S., *Relationship Between Platoon Gunnery Training and Live Fire Performance*, Alexandria, Va.: U.S. Army Research Institute, ARI Research Report 1701, September 1996.

U.S. Army Forces Command, *Training at the National Training Center*, FORSCOM Regulation 350-50-1, Fort McPherson, Ga., January 2010.

U.S. Army Training and Doctrine Command, *Guide for Developing Collective Training Products,* Pamphlet 350-70-1, Fort Monroe, Va.: Training and Doctrine Command, May 2004.

__________, *The United States Army Operating Concept*, Pamphlet 525-3-1, August 2010.

U.S. Army Training Support Center, *Operational Requirements Document (ORD) for the Fire Support Combined Arms Tactical Trainer (FSCATT)*, memorandum, March 1993.

U.S. Department of the Army, *Infantry Live Fire Training*, TC 7-9, September 1993.

__________, *Planning, Programming, Budgeting, and Execution System*, Army Regulation 1-1, Washington, D.C., February 1994.

__________, *Helicopter Gunnery*, FM 3-04.140(FM 1-140), July 2003.

__________, *Army Range and Training Land Strategy*, Washington, D.C., February 2004.

__________, *Approval of the Engagement Skills Trainer, Increment II (EST II) Capability Production Document (CPD),* Memorandum, Washington, D.C., February 2006a.

__________, *Crew Served Machine Guns 5.56-mm and 7.62-mm*, FM 3-22.68, Washington, D.C., July 2006b.

__________, *The Infantry Rifle Platoon and Squad*, FM 3-21.8, March 2007.

__________, *Heavy Brigade Combat Team (HBCT) Gunnery*, FM 3-20.21/MCWP 3-12.2, Washington, D.C., September 2009a.

__________, *Total Army Munitions Requirements Process and Prioritization System*, Army Regulation 5-18, Washington, D.C., December 2009b.

__________, *Index and Description of Army Training Devices,* Washington, D.C., May 2010a.

__________, "Training Ammunition & Simulation VCSA Portfolio Review," briefing, May 2010b.

__________, *Training Ranges*, FM 25-8, Washington, D.C., May 2010c.

__________, *Field Artillery Gunnery (Revised Final Draft)*, TC 3-09.8, Washington, D.C., November 2010d.

__________, *Rifle Marksmanship M16/M4-Series Weapons*, FM 3-22.9, Washington, D.C., February 2011a.

__________, *Army Force Generation*, Army Regulation 525-29, Washington, D.C., March 2011b.

__________, *Avenger Gunnery Program (Initial Draft)*, TC 3-01.63, Washington, D.C., May 2011c.

__________, *Training and Leader Development*, Army Regulation 350-1, Washington, D.C., August 2011d.

__________, *Unified Land Operations*, Army Doctrinal Publication (ADP) 3.0, Washington, D.C., September 2011e.

__________, *2012 Army Posture Statement,* Washington, D.C., February 2012a.

__________, *Infantry Rifle Company Collective Task Publication*, TC 3-21, September 2012b.

U.S. Department of Defense, *DoD Modeling and Simulation (M&S) Glossary*, March 2010.

U.S. Marine Corps, *Marine Corps Combat Marksmanship Programs*, Marine Corps Order 3574.2K, August 2007.